The Logic of
Personal Knowledge

THE LOGIC OF PERSONAL KNOWLEDGE

Essays Presented to
Michael Polanyi
on his Seventieth Birthday
11th March 1961

London
ROUTLEDGE & KEGAN PAUL

*First published 1961
by Routledge & Kegan Paul Limited
Broadway House, 68–74 Carter Lane,
London, E.C.4.*

*Printed in Great Britain
by Cox and Wyman Limited
London, Fakenham and Reading*

*No part of this book may be reproduced
in any form without permission from
the publisher, except for the quotation
of brief passages in criticism*

Q
171
.L84
1961

CONTENTS

Acknowledgements *page* vii

Subscribers ix

PRELUDE

1. The Hungary of Michael Polanyi   PAUL IGNOTUS   3

PART ONE: THE SCIENTIST AS KNOWER

2. An Index to Michael Polanyi's Contributions to Science   JOHN POLANYI   15
3. Polanyi's Contribution to the Physics of Metals   ERICH SCHMID   19
4. Rates of Reaction   HENRY EYRING   25
5. The Size and Shape of Molecules, as a Factor in their Biological Activity   E. D. BERGMANN   37

PART TWO: HISTORICAL PERSPECTIVES

6. Kepler and the Psychology of Discovery   ARTHUR KOESTLER   49
7. The Scientists and the English Civil War   C. V. WEDGWOOD   59
8. Vibrating Strings and Arbitrary Functions   J. R. RAVETZ   71
9. The Controversy on Freedom in Science in the Nineteenth Century   J. R. BAKER   89

PART THREE: THE KNOWLEDGE OF SOCIETY

10. Max Weber and Michael Polanyi   RAYMOND ARON   99
11. Centre and Periphery   EDWARD A. SHILS   117
12. The Republic of Science   BERTRAND DE JOUVENEL   131
13. Machiavelli and the Profanation of Politics   IRVING KRISTOL   143
14. Applied Economics: The Application of What?   ELY DEVONS   155

Contents

15. Some Notes on 'Philosophy of History' and the Problems of Human Society   D. M. MACKINNON   171
16. Law-Courts and Dreams   ELIZABETH SEWELL   179

PART FOUR: THE KNOWLEDGE OF LIVING THINGS

17. The Logic of Biology   MARJORIE GRENE   191
18. Origin of Life on Earth and Elsewhere   MELVIN CALVIN   207
19. The Probability of the Existence of a Self-Reproducing Unit   E. P. WIGNER   231

   Bibliography   239

ACKNOWLEDGEMENTS

The editors wish to thank all the subscribers, and especially the Congress for Cultural Freedom and the Fund for the Republic for their generous support. We are also grateful to all those who helped to organize the volume; we should mention in particular Miss Olive Davies who gave us invaluable clerical assistance.

We wish to thank Messrs. George Allen and Unwin for permission to use Professor Devons's paper, which forms part of his forthcoming volume of essays. The research for Professor Calvin's paper was facilitated by the United States Atomic Energy Commission. Dr. John Robertson of the University of Leeds has kindly assisted in the translation of Professor Schmid's contribution.

In the notes and references throughout this volume, *Personal Knowledge* is designated by the abbreviation *P.K.*

SUBSCRIBERS

The University of Aberdeen
Dr. Franz Alexander
Dr. A. O. Allen
A. H. Allman
Amherst College
Mrs. M. H. Armitage
Dr. John R. Baker
R. W. Baldwin
Sir Thomas Barlow, G.B.E.
Chester J. Barnard
Dr. Julius G. Baron
Prof. M. S. Bartlett
Paul A. Beck
Leonard F. Behrens
Daniel Bell
Prof. Max Beloff
Prof. E. D. Bergmann
Miss Marion Bieber
The University of Birmingham
Prof. P. M. S. Blackett
Mrs. Kathleen Bliss
Dr. Ernest Blumenthal
Dr. B. V. Bowden
Brown University, Providence, R.I.
E. R. Braithwaite
Prof. Dr. Rudolf Brill
The University of Bristol
Bryn Mawr College
Dr. G. N. Burkhardt
Prof. A. K. Cairncross
Dr. E. F. Caldin
Prof. Melvin Calvin
Prof. J. M. Cameron
Dr. A. S. Carson

Signor Alberto Carocci
Professor T. E. Chester
Colin Clark
Dr. Mary G. Clarke
Prof. John Cohen
Dr. J. B. Conant
Congress for Cultural Freedom
Dr. J. W. Cook
A. R. Cooper
Prof. W. Mansfield Cooper
Prof. C. A. Coulson, F.R.S.
Prof. and Mrs. M. Cunliffe
Prof. James R. Curry
Rev. and Mrs. J. F. Davidson
Miss Olive Davies
Prof. S. R. Dennison
E. G. Dentay
G. E. Doan
Prof. D. D. Eley
Prof. Dorothy M. Emmet
Mrs. M. Evans
The University of Exeter
Mrs. K. R. Farrar
Prof. The Rev. T. Fish
Norman G. Fisher
Julius Fleischmann
Dr. Paul J. Flory
Mrs. Elena Frank
Fritz-Haber-Institut der Max-Planck
   Gesellschaft
Prof. Lon. L. Fuller
Fund for The Republic
Prof. D. Gabor
Prof. Max Gluckman

Subscribers

Dr. Brian G. Gowenlock
Dr. Marjorie Grene
Dr. C. E. Griffin
Prof. Gottfried Haberler
Prof. F. C. Happold
Prof. Sir Alister Hardy
Harvard College
Prof. F. A. Hayek
Dr. W. Heller
J. B. M. Herbert
Dr. Mary B. Hesse
Sir Hector Hetherington
Sir Cyril Hinshelwood
Dr. Trygve J. B. Hoff
Dr. and Mrs. G. Hollo
Dr. Julius Hollo
Prof. Juro Horiuti
Dr. C. Horrex
The University of Hull
Hungarian Writers' Association Abroad
Robert M. Hutchins
Walter James
Prof. Sir Geoffrey Jefferson
Prof. John Jewkes
Johns Hopkins University
Prof. H. G. Johnson
M. le Baron de Jouvenel
Andor A. Kemeny
Prof. George F. Kennan
King's College, Newcastle upon Tyne
Prof. Frank H. Knight
Arthur Koestler
Irving Kristol
Melvin J. Lasky
Prof. Arthur Lewis
The London Library
The University of London Library
Prof. W. J. M. Mackenzie
Mrs. Virginia McClam
Dr. Herman F. Mark
Dr. R. A. Markus
Dr. Jacob Marschak
Max-Planck-Gesellschaft z. F. d. W.
Max-Planck Inst. für Phys. Chem.
Geoffrey W. Meadows
Merton College Library
Joseph V. Morello
Sir Charles Morris
J. R. J. Mure
Prof. John U. Nef
Greville Norburn
The University College of North Staffordshire
Dr. J. H. Oldham
Prof. and Mrs. R. A. C. Oliver
Prof. and Mrs. Hugh O'Neill
Prof. Haruo Ootuka
Herbert Passin
Prof. R. Peacock
Niel Pearson
Sir Robert Platt, Bt.
Dr. P. H. Plesch
University of Pennsylvania
Mr. and Mrs. George Polanyi
Prof. and Mrs. John Polanyi
Mrs. M. Polanyi
Prof. George Polya
Prof. W. H. Poteat
Queen's University, Belfast
Dr. J. R. Ravetz
Prof. S. G. Raybould
Sir Charles Renold
Dr. Paul Rosbaud
Dr. and Mrs. S. Rothman
Dr. A. L. Saboz
Mr. and Mrs. W. Schoenheimer
Miss M. Elizabeth Sewell
Prof. Edward Shils
H. A. Siepmann
Rt. Hon. Lord Simon of Wythenshawe
Charles Singer
Prof. C. S. Smith
Prof. R. N. Spann
Mrs. Toni Stolper
Lord Stopford of Fallowfield
Dr. E. B. Strauss
Dr. Michael S. Striker
Prof. D. W. G. Style

Subscribers

Dr. George Szasz
Sir Hugh Taylor
Sir Thomas Taylor
Prof. A. E. Teale
University of Texas
Prof. Sir Alexander Todd
Dr. M. Tyson
Mr. and Mrs. Ruel W. Tyson, Jr.
Prof. F. A. Vick
Prof. J. M. Wallace-Hadrill
Prof. Ross D. Waller
Sir Francis Walshe

Prof. C. W. Wardlaw
Dr. Ernest Warhurst
Miss C. V. Wedgwood
Dr. M. A. Weinberger
Dr. H. Westman
Prof. Eugene P. Wigner
Neal Wood
Sir Hubert and Lady Worthington
Prof. B. A. Wortley
Dr. Hans Zeisel
Mr. and Mrs. W. H. Zimmern
Prof. Sir Solly Zuckermann

PRELUDE

As I acknowledge, in reflecting on the process of discovery, the gap between the evidence and the conclusions which I draw from them, and account for my bridging of this gap in terms of my personal responsibility, so also will I acknowledge that in childhood I have formed my most fundamental beliefs by exercising my native intelligence within the social milieu of a particular place and time. I shall submit to this fact as defining the conditions within which I am called upon to exercise my responsibility. *P.K.*, p. 322–3.

1

The Hungary of Michael Polanyi

PAUL IGNOTUS

THE Hungary of Michael Polányi's youth was a most conservative country cherishing radical ideals. For centuries, Hungarians used to be a nation of warriors, heroes and double-crossers, now fighting, now making accommodations with the armies of the Habsburg and the Ottoman empires who had invaded their country. The Hungarians sometimes hoped to be liberated from the Austrians by the Turks, and sometimes from the Turks by the Austrians; they were astute in playing one conqueror off against the other, and as dignified as losers can be whenever they failed in this game. By the beginning of this century, Turkish domination had already, for all political purposes, fallen into oblivion, but the presence of the Austrians was a reality, psychologically even more than legally. Hungary was then the junior partner in the Habsburgs' Austro–Hungarian Empire, enjoying a home-rule more complete in a way than that of the senior partner but still not completely emancipated from her colonial status, inherited from the period of Turkish and Austrian conquests. For, since 1867, the *Ausgleich* had been in existence between Vienna and Budapest, a compromise unpopular with the Hungarians but with which, at bottom, none of them would have liked to dispense. Under this system, Hungarian foreign affairs, national defence and finances were largely controlled by Vienna, but in exchange the Hungarians—and mainly their landlords, civil servants and county officials—were free to administer a Hungary whose population included, broadly speaking, as many non-Hungarians as Hungarians. This arrangement could hardly be maintained otherwise than by a daily give-and-take between the central power in Vienna and the proud satellite power in the Hungarian Kingdom; and one of the most important Hungarian rights the Emperor had to put up with was the *jus murmurandi* of the Hungarian gentlefolk, a right indulged by them through stormy filibustering, street demonstrations and rhetorical pamphleteering, claiming to perpetuate the spirit of 1848 when Hungary had fought for complete independence and civic rights, and branding the Austrian tyrants who, assisted by the Russian Tsar's army, had crushed that noble manifestation of the freedom-loving national spirit.

Once a compromise was reached with the Court, such clashes, in and outside the legislative bodies, came and went and were followed by periods of relative calm; the government went on ruling in a loyal but uneasy harmony with the Court, and the Opposition went on clamouring for a complete independence, the prospect of which was a nightmare as well as a Utopian dream. The great disenfranchised majority of the population—mainly peasants, farm labourers, farm hands and very poor farmers, Hungarians as well as non-Hungarians—went on tolerating their lot, exhibiting their loyalty when occasion required, murmuring when occasion also required, and hoping in a vague way that ultimately either murmuring or loyalty might help.

All this was bad enough but not so bad as it sounds. It was Merry Old Hungary with all the solace of rituals, songs, alcohol, occasional plenty and permanent self-deception offered by an outdated society to its average members. Its gipsy music has achieved great popularity all over the world, with its intermittent moods of sentimentality, ebullient gaiety and knightly coyness. It was tailor-made for the haughty and frustrated squire who had it played in his ear by the courteous gipsy fiddler. 'And foreigners think *this* is Hungary,' our intellectuals would disapprovingly sigh and then add that it was a sham, the brilliant surface of a sea of darkness, and not even really brilliant, not even good art, not even genuinely Hungarian, not ancient, but modified by the conventions and mannerisms of the ruling classes. 'Isn't it absurd to see Hungary through that?' they would ask. Surely it was absurd; but Hungarians, even more than people outside their frontiers, did see themselves through it. So did the squires, because they liked to behave like squires, and so did the peasants because they also liked to behave like squires. So, even, did the intellectuals when they had drunk a bottle or two of wine and forgotten their principles. The tunes of the gipsy music were the lingua franca of national sentiment.

The ruling idea was liberalism. Practically all parties were devoted to it in theory, and all reproached one another for being so in theory only. No doubt, all were right. With a past and in a position such as that of Hungary, both liberalism and its adulteration seemed unavoidable. Parliamentarism, religious tolerance, a general contempt for prohibitions and restrictions, the will to industrialize the country and to rely on education rather than on authority; all this had been in the tradition of the Hungarian nobility. They had stood up for such ideas in their fight against Habsburg absolutism and by doing so had succeeded in securing a privileged standing for their nation in the Empire and privileged positions for their own sons in the Government and county offices. To destroy that liberal tradition would have been absurd; it would have annihilated the *raison d'être* of the Hungarian rulers. But to take the liberal idea at its face value would have been no less absurd. It would have meant pure Hungarians giving way to non-Hungarian 'nationalities' in what they considered their own country, and the submersion of the gentry in the masses of shopkeepers, artisans, and small farmers, whom, for all their theoretical belief in equal rights for all, they could not help regarding as a different species.

So, more precisely, the era from 1867 to 1918 was that of lip-service to Liberty. And lip-service is a rewarding thing. A class cannot constantly and consistently deceive either itself or others and is, at any rate occasionally, made to keep its word. The Hungary of sham liberalism and qualified independence was a country of fascinating progress. It created cities and factories, model universities, leading engineers, doctors, psycho-analysts, poets and composers. Its legal system was in many ways more humane and indeed more advanced than that of the most advanced countries of those days. Legalism was not only highly developed; it was hypertrophic, dating from times when it had served as a weapon of the noblemen against central power and the new-comers. To what extent were those humane laws actually observed? There was surely more in the cities than in the villages, and more for the gentry than the rest. Some poor crumbs, however, were available to everybody.

Whether for good or evil, there was until the war censorship of 1914, a free press—much freer in libelling and blackmailing than that of many a Western country today, especially the United Kingdom. Such an ignoble variety of freedom is not always separable from a true spirit of imaginative criticism, which was in fact at its height in the Hungary of those days. True, the Socialist daily was more often than not banned from the newsvendors' stands in the streets, and a farm-hand might have risked his head by subscribing to it. Yet the very freedom to print such matter, and material more subversive still, was a national treasure, and a consolation for the country's failure to achieve her own ideals.

The huge palace housing the Parliament, divided into lords and commons, built in the Gothic style on the bank of the Danube, conspicuously on the Westminster pattern but bulkier and far more comfortable than its original, was a theatre of most spirited debates, displaying a high standard of oratory, of legal-mindedness, and of venomous demagogy. Deputies were elected by open ballot and restricted suffrage, which accounted for their interests being limited. They hardly ever touched on the most vital problems, such as the strikingly unequal distribution of landed property, the oppression of peasants and other plebeians by the privileged county bureaucrats and gendarmerie (an oppression even more cruel than that inflicted by their landlords), or, most delicate of all, the fact that the Hungarian nation consisted largely of 'nationalities' which did not wish to belong to it. But there were other matters to discuss, such as the disregard by Austrian magistrates of Hungarian sovereigny, or the recurring demarcation dispute between Church and secular authorities, or, mainly, scandals of racketeering and corruption. The press galleries were so built as to be almost indistinguishable from the rows where legislators sat; indeed, peers and elected commoners, in the lobbies and the buffets no less than in the Houses of Parliament themselves, paid more attention to the reporters than to one another. Austria was at that time, in the words of the wise Socialist leader, Dr. Victor Adler, a *durch Schlamperei mildeter Absolutismus*, an imperial absolutism slipshod and easy-going enough to make life bearable. Hungary, it may be added, was an oligarch *democratique* by the hunt for publicity.

There were two layers of society, or rather two networks of social groups, consistently opposed to this state of affairs. One was Social Democracy. It stemmed from 1848. The first champion of a socialist workers' movement was a self-educated proletarian writer of those days, touchingly self-devoted, prolific, and tiresome, a heroic inmate of the Habsburgs' jails and, after the *Ausgleich*, member of parliament. He believed in the brotherhood of men and in the destiny of Hungary to assimilate them all. Of course Hungary was not the only country where revolutionary socialist sentiments had been cradled simultaneously by Messianic internationalism and intolerant nationalism. But in the 1900s this was to some extent past history. Social-Democracy (Party and Trade Union) had by then acquired the position of the most important organization expressing a desire for far-reaching social change. Its main supporters came from skilled labour in big cities. It had numerous adherents among white-collar workers, office clerks, and shop assistants, as well as among manual workers of the poorest categories, such as bricklayers, miners, and navvies. In the market towns, where police and gendarme terror were less omnipotent than in rural areas, there were also some hired farm labourers and tenant farmers in its ranks.

But the archetype of the Social Democrat was, I should say, a cross between a carpenter and a compositor. I remember May-Day meetings where I lingered as a romantic grammar-school boy, hoping to discover the apache types I had seen on the stage as the personifiers of Robespierre's mob. What a disappointment it was to see instead carefully dressed lower-middle-class types, with brilliant red neck ties but preferably wearing bowler hats. There was a middle-aged, short gentleman, with a big, well-groomed moustache, holding the huge red banner, and now and then shouting rhythmically 'LONG LIVE WORLD-REDEEMING, INTERNATIONAL, REVOLUTIONARY SOCIAL DEMOCRACY!' Socialism in Hungary was based on Marxism in theory, and concentrated on *Sozialpolitik* in practice. It was the suicidal narrow-mindedness of the ruling gentry that prevented, through the existing 'electoral' system, such decent 'Godless and country-less scoundrels' from being represented either in the legislative bodies or, as far as I remember, in any of the local government councils.

There are moments of chaos and convulsion in history when pauperdom speaks for itself. As a rule this is not the case; the constant urge for radical changes on behalf of the have-not's comes from those who 'have', though not enough. The comparatively best-paid layer of workers was thus one of the hotbeds of social discontent. The other source was, in an even more clamorous way, that part of the white-collar class not completely accepted as gentlemanly. With some over-simplification I may say that this was the Jewish middle class.

The social composition of the country accounted for this fact. There was a greater proportion of lesser nobility in Hungary than anywhere else in Europe. Correspondingly it was poor; but the poorer it became, the more stubbornly it stuck to its privileges—until 1848, freedom from taxation and inalienability of landed propery,* and after the declaration of civic rights for all, a *de facto* monopoly over the county administration and other posts in the public service. As a result, the dispossessed landed gentry and the Magyarized offshoots of the ancient German guild *bourgeoisie*, who outdid the blue-blooded in social snobbery, had got used to living on the assumption that any job not based on the monopoly of land or of the public administration was unworthy of a gentleman. In the meantime, it was a foregone conclusion that free enterprise must be encouraged, in commerce and banking as well as in industry. Who, then, should act as *entrepreneurs* and who should deal with the trades and professions rooted in the new world of free enterprise? The noblemen and their adopted ex-German brethren were too lazy or too dignified to do this and the rural-proletarian bulk of the population was too miserable even to think of it. So this role automatically fell to Jewry. The Jews were later reproached for 'taking possession' of trade and banking. This is as correct to say as that the outcaste Hindus took possession of the jobs despised by the caste Hindus. On the other hand, of course, the 'untouchability' of the Jews in Hungary provided a much more lucrative status than that of the pariahs in India.

Thus, from the end of the last century onwards, two middle-classes lived side-by-side in Hungary: the 'historical middle-class', which was a euphemism for those enjoying inherited privileges, and the '*bourgeois*' (*polgári, bürgerlich*) middle-class, which comprised:

(a) artisans of the kind referred to as 'semi-proletarian' in Marxist literature, and, more important,
(b) capitalists and merchants, as well as professional people and other intellectuals, of Jewish religion or Jewish extraction.

Perhaps the saddest aspect of this development was the *pernicious anaemia* of the spirit of the ancient Hungarian nobility. Heirs of an intellectual *élite* second to none in broadmindedness and generosity, who in the 'thirties and 'forties of the last century had wedded Hungary to liberalism, they had later grown rigid and made a virtue of dandified ignorance. There were still outstanding personalities among them who

---

* A great proportion of the magnates' and bishops' latifundia went on being entailed property until 1945 when all latifundia were done away with. But such entailments were not, after 1848, extended to smaller-sized properties.

found this atmosphere unbearable; but these had to desert their own hereditary environment and to take refuge in the '*bourgeois* middle-classes'. This, to put it once again less euphemistically, meant marrying Jewesses or visiting their *salons*.

The '*bourgeois* middle-class' itself, reacted in three different ways to this distribution of social functions. One was the same as that of the Magyarized German guild-*bourgeois*—to outdo the genuine squire in duelling, horsemanship, gipsy-serenading and all he was supposed to stand for. The second was wisely to accept the existing caste system and to make the best of an arrangement which provided the privileged with the most decorative visiting cards, titles, and coats-of-arms, but allowed the under-privileged tradesman to make more money. The third pattern of behaviour was protest mingled with artistic and scientific curiosity. The well-to-do underdog is the most embittered of all; even if refined and exhilarated by the taste for pleasure and learning, he keeps those seeds of prophetic grief and fury which, judged by the crops they yield, may turn out either noble and creative or devastatingly vulgar or both. Was it the huge and indeed incessant record of suffering of the Jews that sensitized them to every sort of intellectual curiosity? Or was it their intellectual aptitude, traceable back to their occupations in the Middle Ages, which taught them to react so intensely not only to their own sufferings but also to those of others? Whatever the psychological background, it is a fact that a substantial minority of the *bourgeois* middle-class in the Hungary of the decade preceding the First World War provided an admirably fertile soil for art and letters, for experiments and speculations, and for dreams of the shape of things to come.

This was the heyday of Hungarian radicalism. Not that the Hungarian Government or even its official opposition had much to do with it. Radicalism, like Social Democracy, was unrepresented in parliament, and both were further from power than was the popular front in the Britain of the late 'thirties; they did not have even a Sir Stafford Cripps to be expelled from a parliamentary party as a punishment for associating himself with them. All the same, they were the pacemakers of Hungary. They were banned from high offices and exclusive clubs; neither the aristocracy, the clergy, the high bureaucracy, the landed gentry, nor the conformist bulk of the capitalists and *bourgeois* middle-classes, Jews and Gentiles alike, would have allowed them to play a decisive part in running public affairs; but at the same time they all instinctively regarded them as the men of the future. Radical intellectuals were surrounded with both contempt and awe. Their influence was enormous; they infiltrated the powerful popular press; but their chief organs were two highbrow periodicals, one literary, one sociological.

The literary fortnightly *West* (*Nyugat*) was productive of much creative genius and reflected cosmopolitan responses to all modern trends. The 'West' that it cherished with most longing was represented by Paris and

Florence; but it was Vienna, Zurich, Munich, Berlin—the Berlin of the modern stage and not that of the Siegessäule—that in fact influenced them most. Attached to Latin elegance through nostalgia and to German *Kultur* through links determined by geography, it embraced greedily every experiment in style and thinking. But it was far from just aping the West. Its writers and artists were too original to do so even if this had been their wish. But it was not; their hankering after the Western Cosmopolis was qualified by their awareness that they were perpetuating ancient Eastern features of character, inherited through language, melodies, images, and perhaps even blood. The leading poet of *West*, Endre Ady, who came from the Calvinist landed gentry, was to some extent an interpreter of modern French literary and political ideas, but for the most part recalled ancient Hungarian passions and visions, rich with biblical images. There was something symbolic in the fact that the composers, Bartók and Kodály, first received publicity for their new music, inspired by ancient folk-lore, through *Nyugat matinées*, and that the magazine carried for many years a somewhat cubicized peasant embroidery on its front page. *West* stood for the absolute freedom of the spirit, including its freedom to scorn freedom. As a group it was politically uncommitted, but the fact that it claimed the right to be so implied a challenge to the belief in ruling ideas and authorities. It was 'Lib-Lab' by the strength of its taste and interests, a pillar of the then extra-parliamentary Left.

The periodical for political science, *Twentieth Century* (*Huszadik Század*), was more radical in its views and less so in its taste. Its editor was Professor Oszkár Jászi, later the founder of the Citizens' Radical Party. It had a profound reverence for a factual approach and for evolutionary theories; which did not, in practice, prevent it from becoming a platform for visionary enthusiasts. The power of literature over science, and that of rhetoric over literature, were so strong in that Hungary which was attached for good or ill to the memories of 1848 that even statistical analyses would have struck their readers as meaningless without a trumpet-like conclusion. Professor Jászi had, both in his character and in his sympathies, much in common with the upholders of English non-conformism. His radicalism consisted mainly in taking the liberal doctrines at their face value. The idea with which he shocked the ruling authorities and made himself a political outcast happened to be just the one devised as a means to save traditional values, those embodied in the Habsburg Monarchy and in historical Hungary. This was the idea that the multi-national character of the Danubian valley should be recognized by transforming it into an 'Eastern Switzerland', a federation of peoples under a constitutional monarch. It was common sense; it would have been the only way of saving all the peoples from wars, massacres, tyrannies, alternate waves of oppression from outside, and madness from inside. It was so much common sense as to prove a quixotic dream.

What was the essence of Hungarian radicalism? It was an attitude of general receptivity rather than a clear-cut programme. A group of intellectuals (including Ady), with Jászi as its political leader, was committed to a programme very similar to that of the French Radical-Socialist of those days, with Fabian leanings. The intellectual leadership of Social Democracy, as well as its most distinguished heretics, professed Marxism. In general, however, the artists, authors, scholars, and scientists, active in and round *West* and *Twentieth Century*, held very motley and conflicting views. While one adhered to absolute liberalism as a disciple of Henry George, another believed, like Sorel, that progress required the disappearance not only of liberalism but of the respect for liberty altogether. While one swore by Haeckel and had only sneers for any idea not strictly fitting in with a ready materialist pattern, another triumphantly discovered Bergson's philosophy as the knell of anything smacking of materialism. An esoteric pride in aristocratic seclusion was no less characteristic of these intellectual circles than was an enthusiasm for the holy illiteracy of the Coolies of the World and for their right to govern the rest of mankind. Let me just mention two personalities conspicuous for their contempt of '*bourgeois* freedom' at that time; one, a believer in aristocratic seclusion, and another, a zealot for mass instincts and for something like a socialist revolution that he saw approaching. The former, a young philosopher, marching in the footsteps of German idealists, George Lukács by name, emerged some years later, on the eve of the Hungarian Soviet venture, as the leading revolutionary Marxist theoretician of his country; the latter, a dithyrambic left-wing philologist and novelist, called Dezsö Szabó, emerged later, after the collapse of the Hungarian Soviet venture, as the idol of Hungarian counter-revolutionary youth and as a codifier of a Hungarian version of *Blut and Boden*.* The strange thing is that neither of them had really much to modify in their styles when undergoing this enormous change. Nor had indeed the camps of their admirers.

The readers' camp of *West* and *Twentieth Century*—the Hungarian intellectuals of those days—was as a whole more revolutionary, more progressive, more opposed to all sorts of traditional standards or limitations than were the creative artists and thinkers it admired. Among the latter, many who crusaded against tradition in one field were keen on vindicating it in another. It was their flock—a flock comparatively small in number and quite well-to-do on the average, but still a flock—which on all occasions backed the anti-tradition crusaders. It was Bloomsbury-on-Danube, at its best and worst.

---

* To do them justice, this applies to their attitudes in 1919 and the early 'twenties. Later, and particularly after the initial successes of German National Socialism, in reaction to the horrors and stupidities enforced under Hitler's and Stalin's tyrannies, they both became more broadminded and democratic in outlook. Dezsö Szabó was a naïve Germanophobe who fought German imperialism bravely and never realized that he was spreading its ideology in a puerile 'Turanian' guise. He died, in tragic circumstances, in the siege of Budapest, 1945. As to Lukács, his career in the last fifteen years has been fairly well known; in any case he is rightly regarded as one of the most important intellects who have tried to reconcile Lenin's interpretation of the Marxist creed with the humanist European heritage.

The Hungary of Michael Polanyi

The most active part of that intellectual group was the Galileo Circle (Galilei Kör) of progressive undergraduates, founded in 1908. At the time of its foundation the existing undergraduates' associations were either denominational in character or emphatically political, particularly nationalist. The cry of the 'Galileists' was to concentrate on science instead. They meant this honestly, passionately, too passionately perhaps. The man who had initiated the circle and who, second to Ady, had the greatest authority with its audiences, the legal philosopher, economist and physiologist, Professor Julius Pikler, was anything but an extremist. He was uncompromising about science, and what made his attitude provocative in politics was nothing more than his expressed conviction that scientific research and speculation must never give way to religious, social or political considerations, however lofty these be. He warned his followers against going further than that in provoking conventional feelings. He warned them in vain. Within a year or two the Galileo Circle outstripped its rival associations in displaying political passions and in keeping the question of the existence of God on its daily agenda. It had planned a Hungarian society that was to follow doctors, mathematicians, and physicists rather than the lawyers who had hitherto swamped the national culture. It had dreamt of a lingua franca of equations, functions, figures, diagrams to substitute for gipsy music and for political sloganizing; but its manners belied its claims. Teeming with Eton-cropped girls and Montmartre-styled boys, it created a platform where the creed of being matter-of-fact was preached with the oratory of the Catacombs. If *West* was permeated with a synthesis of Latin *ésprit* and Teutonic *Kultur*, and *Twentieth Century* with an Anglo-Saxon ideal of fairness, the Galileo Circle reflected the spirit of revolutionary Russia, rationalism driven to its mystic extreme.

The first president of the Galileo Circle was a fascinating young sociologist, sophisticated, witty, eloquent; Karl Polányi, a Professor emeritus at Columbia University today. When in the heat of a brawl at the University he was challenged to a duel, he answered: 'I'm always pleased to fight you by intellectual arms.' It was, as far as I know, this piece of repartee that had established his popularity: I heard it quoted again and again. I am inclined to pay him the tribute of calling him the most left-handed politician I have ever come across. He may always have been dominated by a fear of lagging behind the times, a fear capable of driving one into such bold extremes one day that one appears outdated the day after. Sound judgement was not, on the whole, the strength of the Galileo Circle. But it was a place able and ready to appreciate all sound ideas—those striking the average Hungarian as sheer eccentricities at that time, and accepted as truisms today.

This was more or less the doom of the whole radical movement of pre-First-World-War Hungary. Its members were most sagacious prophets, but apart from creating works of art and thought—even occasional masterpieces—they achieved nothing. If Hungary and Eastern Europe have advanced in many ways since then, as they have receded in many others, this has come about, apparently, under the pressure of world events, independently of what had first been clarified, in an unusual but most expressive language, on the pages of a *Nyugat* that meant West, and a *Huszadik Század* that meant Twentieth Century, and on the platform of the Galileo Circle. Surely, by paying more attention in time to those radical warnings, great values and millions of people could have been saved from disaster. Would this have been possible at all? Opinons may differ on that. But for one interested in the development of ideas from the state of embryonic visions to their consolidation in pedestrian reality, the intellectual history of Hungary in the first two decades of our century reveals extremely rich material.

I should like to dedicate these recollections to Michael Polányi. They were meant to give an idea of the surroundings in which his career started. Since the beginning of the inter-war period, he has no longer lived and worked in Hungary. Yet I have always had the impression that he has remained a citizen of that radical republic of letters—partly through loyalty to its unwritten constitution and partly through reacting against it. His family was one of intellectual brilliance and playful political nonconformity. So was the circle of his friends whom I met. His reputation was that of the man who had the courage to dissent from the dissenters; in a flock of black sheep he shocked many by seeming almost white. He was a founder member of the Galileo Circle. Today he is a great scientist, opposed to scientism. I do not know whether this was already his outlook fifty or fifty-two years ago, among his fellow-members who believed in science above all. What we know is that while others excelled in extolling science, he excelled in practising it. Like his brother, Karl, he has remained faithful to radicalism. But unlike his brother, the most radical of radicals, he was the most moderate of radicals. Radicalism in the search for truth has led him to the re-discovery of basic tenets which no liberal movement or liberal profession can disregard without destroying its own foundations.

I cannot help feeling that the intellectual environment of his youth has profoundly influenced his development. From it he inherited the limitless liberality of his mind, the simultaneity of personal and technical interests, and the ability to co-ordinate them in behaviour as well as in philosophy. What made him differ most from those around him was his reverence. He thinks that the intellectual youth of Hungary and of some other countries, when 'revising' its dogmas of revolutionary origin, is meeting him on the grounds of re-discovered ethical traditions. Whether or not this applies in every respect to the young men of our time, it does apply to Michael Polányi himself. The inherent radicalism and the scientific sensibility of his intellect have made him up to date, and the grain of conservatism in it, his attachment to the perennial, enables him to be ahead of his time.

PART ONE

The Scientist as Knower

Scientists . . . spend their lives in trying to guess right. They are sustained and guided therein by their heuristic passion. We call their work creative because it changes the world as we see it, by deepening our understanding of it. The change is irrevocable. A problem that I have once solved can no longer puzzle me; I cannot guess what I already know. Having made a discovery, I shall never see the world again as before. My eyes have become different; I have made myself into a person seeing and thinking differently. I have crossed a gap, the heuristic gap which lies between problem and discovery. *P.K.*, p. 143

2

An Index to Michael Polanyi's Contributions to Science

JOHN POLANYI

My father's scientific interests have taken him into a substantially larger number of independent fields of research than is customary. This makes it difficult to follow his trail. A scientific bloodhound can follow him through a particular field. But today the well-adjusted hound is a specialist; if the quarry is an eclectic the scent is rapidly lost. It is hoped that the following summary, which lists his main contributions under four headings, will serve as a partial index to the scientific bibliography given later. Since the names of collaborators are to be found in the bibliography, to which cross references are given, these names have not always been repeated in the summary. I hope that this discourtesy will be forgiven.

Adsorption

1. A theory of adsorption (ref. 12, 1916).
   This theory was critically received for the first thirty-five years after its publication, but has recently come into favour.

2. The application, made jointly with F. London (ref. 115, 1930), of the theory of dispersion forces to the additivity rules of the adsorption potential. These additivity rules had previously been postulated empirically by M.P.

3. Experimental observations made in the light of the theory (1., above) on the state of adsorbed vapours and crystalline solutes (refs. 92, 94, 1928).

4. Observation of very high heats of adsorption at low pressures (ref. 93, 1928). (Paralleled by H. S. Taylor and by others.)

John Polanyi

Plasticity and Strength of Materials

1. Discovery of general hardening of single crystals (ref. 41, 1922).
2. Elucidation, following on experiments by Andrade, of the mechanism of gliding in metal crystals, particularly zinc (ref. 46, 1922).
3. Elucidation of the mechanism for tin crystals (ref. 52, 1923).
4. Discovery of hardening of single crystals against (a) gliding (ref. 48, 1922; 50, 1923; 62, 1924). (b) brittle rupture (ref. 60, 1924).
5. Discovery of 'recovery' in crystals (Refs. as in 4 (a)).
6. Discovery of the reversibility of the tendency towards recrystallization (ref. 105, 1928; 121, 1931).
7. Discovery of 'fibre-structure' in cold worked metal wires (refs. 35, 38, 40, 1921).
8. Determination of the structure for various classes of metals (refs. 35, 38, 1921).
9. Discovery of 'recrystallization-structure' (refs. 38, 1921).
10. First observation of the plasticity of crystals at low temperatures, down to liquid helium temperatures (ref. 113, 1930).
11. 'Dislocation' as an explanation of crystal plasticity (ref. 156, 1934). (Paralleled by G. I. Taylor.)

X-ray Analysis

1. Discovery (with Herzog) of 'fibre structure' in cellulose (ref. 27, 1920). (Paralleled by Scherrer.)
2. Elucidation of the 'fibre diagram' and determination of the elementary cell of cellulose; elimination of all but two types of chemical structures for cellulose (ref. 32, 1921).
3. Establishment (after preliminary experiments by Schiebold) of the 'rotating crystal method' for X-ray analysis of crystals. Discovery, in particular, of the 'layer line relationships' (ref. 45, 1922).
4. Determination by this method (3, above) of the structure of white tin (ref. 52, 1923).
5. Determination, with Brill, of the cell of silk protein (K. H. Meyer, 'Natural and Synthetic High Polymers' p. 446).

Reaction Mechanism

1. First theoretical explanation of an activation energy ($H_2 + Br_2$ reaction). (refs. 18, 19, 20, 21, 1920). (Paralleled by Herzfeld and Christiansen.)
2. (a) Theory of surface dissociation and consequent reduction of activation energy on solid catalysts (ref 34, 1921).
   (b) The same conception expanded in terms of quantum mechanics (refs. 108, 1929; 124, 1932).

3. Rate of activation of a gas-reaction by collisions, calculated for the first time by the 'many degrees of freedom formula' now commonly applied to this problem (ref. 23, 1920).
4. Jointly with E. Wigner, the first formulation for a chemical molecule of the inverse relationship between stability and breadth of quantum state, which is now the basis of pre-dissociation theory. The same paper contains the postulate of Uncertainty for angular momentum (ref. 70, 1925).
5. Jointly with E. Wigner, the theory of the absolute rate of monomolecular reactions (ref. 97, 1928).
6. Jointly with H. Eyring, the calculation of activation energies and reaction mechanism by the energy surface method (ref. 118, 1931).
7. Atomic Reactions of Na Vapour.
    (a) Invention of the highly dilute flame method (with Beutler and Bogdandy). Observations by this method include (i) measurement of the rates of a number of reactions occurring at every collision; (ii) discovery of a luminescence with a high yield and first elucidation of the precise atomic process of a chemiluminescence. (Sodium + chlorine luminescence had been discovered by Haber at a yield of perhaps 10^{-6} *quanta* per reacting molecule.)
    (b) Invention of the 'diffusion flame' method (with v. Hartel). Observations by this method include rates of reaction of sodium with numerous organic halides.
    (c) Invention of the 'life period' method (with L. Frommer) extended the range of measurable reactions. (References to this work are reviewed in ref. 131, 1932.)
8. The mechanism of negative substitution including the accompanying optical inversion (ref. 127, 1932). First kinetic analysis of racemization by atomic exchange (jointly with E. Bergmann) (ref. 136, 1933).
9. Racemization through electrolytic dissociation (optical instability of carbonium ion) first postulated and measured (ref. 142, 1933).
10. Jointly with Horiuti, the first observation of catalytic exchange reaction (water and deuterium on platinum) (ref. 145, 1933).
11. Jointly with Horiuti, the observation of catalysed exchange of deuterium with benzene; theory of the Half Hydrogenated State to explain exchange, hydrogenation and catalytic migration of double bonds (ref. 153, 1934).
12. First use of heavy oxygen for establishment of reaction mechanism (hydrolysis), with Szabo (ref. 150, 1934).
13. Method for calculating the activation energy of an ionic substitution reaction and of the reaction of sodium with organic halides; first derivation of parallelism between reaction heat and reaction rate of a 'homologous' series of reactions (ref. 168, 1935). Later elaborated with M. G. Evans.

14. With Horiuti, the theory of activation energy for prototropic reactions including cathodic discharge (ref. 176, 1935).
15. The above was later elaborated with M. G. Evans into a theory of Bronsted's logarithmic linear relationship between equilibrium and rate constants (ref. 179, 1936).
16. With M. G. Evans, the discovery of a linear relationship between heat of solution and entropy of solution (ref. 179, 1936).
17. Jointly with M. G. Evans (ref. 169, 1935).
    (a) A generalized form of Wigner's 'Transition State Method' for calculating reaction rates.
    (b) The thermodynamics of the Transition State in varying media.
    (c) The postulate of intermediate position (between initial and final states) of transition state in regard to density and entropy.
    (d) The theory of the effect of hydrostatic pressure on reaction rate.
    In this list (a) was paralleled independently in a paper published earlier by Eyring; (b) was paralleled independently in a paper published later by Eyring and Wynne Jones.
18. Systematic measurements, conducted jointly with E. T. Butler, of the rates of pyrolysis of organic iodides have revealed a close parallelism of these with the corresponding rates of the sodium reaction. From this evidence (and other data) considerable variations of bond energy of saturated hydrocarbons were derived. These variations were theoretically explained and related to other properties. Part of this work was done jointly with E. C. Baughan, other parts jointly with E. C. Baughan and M. G. Evans (refs. 198, 199, 1940; 200, 1941; 205, 1943).
19. Investigation (with A. G. Evans and H. A. Skinner) of the mechanism of Friedel-Craft catalysed isobutene polymerization; the notion of 'co-catalysis' (refs 208, 210, 1946; 211, 212, 214, 215, 1947).

3

Polanyi's Contribution to the Physics of Metals

ERICH SCHMID

METALLURGY is of particular importance today, above all because of the great significance for many branches of physics and technology of problems involving raw materials. Thanks to the co-operation of many outstanding workers it has entered in the last few decades upon a very rapid course of development, which is by no means finished. In the following pages I shall try to assess the work and significance of a man who was active in this field for only a few years and with limited experimental facilities, but who has left such clear traces that he must be numbered among the great pioneers of this science.

The road by which Polanyi came to metallurgical research is by no means the usual one. In the year 1920, R. O. Herzog and Scherrer[1] had observed that the cellulose in plant fibres has a crystalline structure. By irradiation of a parallel bundle of fibres, using a perpendicular monochromatic X-ray beam, they obtained diagrams which did not exhibit uniform density of the Debye-Scherrer circles, and could not therefore be explained by random orientation of the micro-crystallites. The circles contained maxima of density only at particular points (4- and 2-point systems), the position of which did not change, even when the bundle was rotated about its vertical axis. If, on the other hand, the bundle was irradiated parallel to its axis, the uniform density characteristic of a micro-crystalline structure with random distribution was obtained. The interpretation of these 'fibre diagrams' in terms of the partial parallel orientation of the crystallites, and the reason for the occurrence of this orientation: these were the problems which Polanyi had undertaken to investigate at Herzog's suggestion—an investigation which he pursued with great originality as leader of a small group of enthusiastic collaborators at the Kaiser-Wilhelm Institute for Fibre Research in Berlin-Dahlem.

The theory of the fibre diagrams was first developed by Polanyi alone, and in its subsequent elaboration together with K. Weissenberg.[2] This theory makes it possible to infer from the distribution of the intensities observed in the diagram, the structure of the irradiated sample. In the case of a simple fibre structure, where all the crystallites lie in the same crystallographic direction parallel with the axis of the fibre, very simple diagrams should result. The interference spectra are confined to layer lines, and from the distance apart of these lines the periodicity of the lattice along the fibre axis can be directly determined. The 'layer line relation' is thus the core of the 'method of crystal rotation' which was developed as a direct result of the interpretation of the fibre diagrams, and which represents one of the most important methods for the determination of crystal structure. The measurement of the layer line distances leads directly to the measurement of the identity period along the axis of rotation of the crystal.

The question of the origin of the structure of such a fibre was first investigated by Polanyi in collaboration with Herzog, Becker and Jancke.[3] The attempt was made to produce such structures by external action on powdered crystals. In particular, compression under high pressure and cold drawing of metal wires, as demonstrated in further experiments with Ettisch and Weissenberg, produced perceptible results.[4] There could be no doubt, that in this case a plastic flow of the grains of the crystal was the cause of the formation of the texture. This conclusion then led on directly to the investigation of the plactic flow of crystals. Since a specially high plasticity was known to exist in metals, the Kaiser Wilhelm Institute for Fibre Research undertook the task, apparently so alien to it, of studying the plastic flow of metallic crystals. Single crystals in the form of wires of various metals with low melting points were chosen as experimental material. These were at first produced by von Gomperz according to the Czochralski method.[5] Along with brittle crystals some crystals susceptible of elongation were also obtained, which permitted flow up to six times their original length. The process of elongation in hexagonal zinc crystals was systematically investigated with the collaboration of Mark and Schmid.[6] By taking account of the translation along the basal plane, the importance of which was recognized in these experiments, and for which the now familiar model was constructed, it was possible to explain the band formation of originally cylindrical crystals, the relation between the degree of elongation and the rotation of the elements of translation with respect to the vertical axis, the end state which the orientation in the widely stretched band was approaching, the stripes appearing on the surface of the band, and many other details. These experiments also demonstrated the low shearing strength of the slip plane and the hardening during elongation. Brittle crystals exhibit orientations with a large angle between the axis of the wire and the basal plane, which appears in such cases as a plane of cleavage.

The external conditions for this work were at first scarcely favourable.

Polanyi's Contribution to the Physics of Metals

The Institute had no building of its own, but was quartered for the most part in the Kaiser Wilhelm Institute for Physical Chemistry. The X-ray laboratory was in the basement of the house of Councillor Haber, the microscopic work was carried on in a room lent by the Kaiser Wilhelm Institute for Biology, and photographic equipment and a metallurgical microscope were available in the government testing office in Berlin Dahlem.

For an extension of the experiments on tin crystals the lattice and translation elements of tetragonal tin had first to be determined. This was done in collaboration with Mark.[7] In this connection also, the dynamics of translation was studied in collaboration with Schmid.[8] For precise and, as far as possible, mutually independent measurement of stress and strain as it proceeds, an apparatus was designed in which the elongation actuates a micrometer screw, and the tension is measured as the sag of a steel spring of appropriate dimensions.[9] This apparatus of Polanyi's is still widely used today both in its original and in modified form. In the tin crystal, again, extremely low shearing strengths of the translation systems were observed, and also an increasing tensile strength with increasing slip. The phenomena of softening in crystals which had been plastically deformed were also extensively studied for this material. These investigations were concerned, firstly, with the recrystallization by the formation of new grains, and secondly, with a process opposite to hardening which occurs with the retention of the lattice position and is dependent on temperature, and for which the name 'recovery' was coined. This phenomenon had previously been known only as the decrease in tensile strength of tungsten crystals under appropriate heat treatment. The experiments with tin crystals demonstrated that even the shearing strength of slip systems in hardened crystals can be extensively modified, in dependence on the temperature and time of heating. Thus recovery affects the shearing strength of slip systems as well as the normal strength of planes of cleavage. An investigation carried out in co-operation with Masing[10] was concerned with the tensile strength of polycrystalline zinc in the compact and heated state and with the dependence of this strength on the size of the metal grains.

Plastic flow experiments on crystals of zinc and tin under hydrostatic pressure (up to forty atmospheres), conducted in collaboration with Schmid, showed that at least for the range of pressure investigated there is no influence of the normal tension on the onset of plasticity.[11]

In several investigations Polanyi studied the low actual shearing strength of slip systems, and the low actual normal strengths of planes of cleavage. Again and again he pointed out that the discrepancy of as much as three or four orders of magnitude between the actual values and those theoretically to be expected cannot be dependent on temperature.[12] And since physical chemistry had long been his chief interest, he examined in collaboration with Meissner and Schmid[13] plastic flow at very low temperatures. In confirmation of his view, he found almost the same low shearing strengths of the basal plane for zinc and cadmium crystals at 1° K. as at room temperature. This behaviour stands in sharp contrast to that of amorphous substances, the strengths of which exhibit a very high dependence on temperature. Several years later, as we shall see presently, Polanyi was to give a decisive impetus to our theoretical understanding of the plasticity of crystals.

For the moment, however, let us recall the work which he carried out in the Kaiser Wilhelm Institute for Metal Research in Berlin-Dahlem. The director of this institute, W. v. Moellendorff, soon recognized Polanyi's achievements as a pioneer in metallurgy, and called on him for advice. In the person of G. Sachs, who joined the institute, he found an extremely gifted co-worker, with whom he conducted some ingenious experiments on inner stresses.[14] In twisted copper rods heated to a moderate temperature, twisting in the reverse sense was observed, which decreased by about 10 per cent with increased temperature. A further increase of temperature, however, produced no effect. This behaviour could be interpreted as the result of recrystallization in various layers of the rod, which was thought of as built up of concentric cylinders. This process will set in sooner in the strongly deformed outer layer than in the inner part of the rod. By using rock-salt crystals as a model substance, it was possible, because of movements involved in the cleavage of the surface layers, to make the stresses remaining in bent crystals directly visible. Experiments on the plastic flow of crystals, with alteration of the direction of flow, produced interesting results. In collaboration with Beck[15] Polanyi succeeded in showing that when a bent aluminium crystal is bent back again, its power of recrystallization is lost, but that on the other hand its hardness increases. In experiments on bent rock-salt crystals conducted in collaboration with Ewald[16] it was discovered that for very slight changes of form the hardening is obtained only when the crystals are bent in one particular direction. The process does not make it more difficult to bend the crystals back; in fact such bending removes the hardening previously effected. This result demonstrates the existence of a purely mechanical process of recovery. Attempts, in collaboration with Schmid, to discover the phenomenon in metal crystals also were unsuccessful at that time. Quickly-stretched tin crystals, which were then slowly stretched still further, did not exhibit plastic flow as a result of this treatment. The turbulence of the flow produced by the elongation and simultaneous activation of several translation systems obstructed the smoothening and weakening activity of a single slip system in the subsequent slow elongation. Today mechanical recovery is effected in metallic crystals as well, in various ways. The significance of the crystal surface for hardness and plasticity was investigated in experiments on rock-salt crystals under water (in collaboration with Ewald[17]), and it was shown that contact with water lowers the threshold of plasticity, and that with drying or the use of saturated NaCl solution this effect is absent. Continual loosening of the surface layer is believed to remove obstacles to the activation of translation and thus to produce the conditions for plastic flow and therefore also for hardening. Elongation of tin crystals in dilute acids produced no analogous effect; but since then results on the significance of adsorption layers for the plasticity of metals have become available.

In a short note in 1935 on a kind of lattice disturbance which was able to produce plasticity, Polanyi gave the key to the interpretation of the low shearing strength of translation systems.[18] It is basically a question of the same model to which he had repeatedly called attention in his discussion of the transaltion mechanism connected with the bending of the slip plane. On the two sides of an activated slip plane, within some definite regions of the plane, n lattice points stand on one side, in the direction of slip, and $n + 1$ opposite. This model had been applied earlier to the interpretation of the hardening of latent slip systems cutting through the actual slip system. If a crystal with such a missing place, which Polanyi called a dislocation, and which is today called an edge dislocation, is subjected in the direction of translation to a shear of a fraction $1/n$ of the lattice period, then the pattern of the dislocation will not have altered, but it will have moved one lattice period away. With reference to the discrepancy of about three powers of ten between actual and theoretical shearing strength, it appears that in metal crystals, which are so easily susceptible of plastic deformation, there must be dislocation points approximately per every thousand atoms. When a dislocation has reached the surface of the crystal, it has exhausted its effects: a slip of one lattice period has occurred. If the capacity for plastic flow is to be maintained, a recurrence of such dislocation must be constantly taking place. This theory of dislocation was suggested at the same time and independently by Taylor.[19] Since then J. M. Burgers, Mott, and many others[20] have used the theory as a foundation for their work and today exact experimental proofs have been produced for it. A glance at current journals on solid state physics and metallurgy will show how fruitful it has proved. Scarcely an issue appears without a number of papers dealing theoretically or experimentally with dislocations and their effects.

In addition to the papers dealing with special questions, Polanyi also considered in a number of clear and fundamental articles and lectures such problems as the structure of matter, the plastic flow of crystals and the plastically deformed state, and the nature of the phenomenon of cleavage. The report of Polanyi and Masing[21] on 'Cold-working and Hardening' in the second volume of the *Ergebnisse der exakten Naturwissenschaften* (1923) made an especially deep impression. A monograph on crystal plasticity, which was planned in collaboration with Schmid, was unfortunately never written. Polanyi's interests had already moved too far into the field of physical chemistry, so that he could no longer undertake the considerable labour which would have been necessary for such a project.

If we survey Polanyi's work on problems of metallurgy, which lasted only a few years, we can see at once that he opened a broad avenue for the exact treatment of a series of highly important problems. The technique and theory of rotation diagrams and of X-ray fibre diagrams, crystallographic and technological characteristics of the primitive translation of crystals, the athermic nature of crystal plasticity, hardening and recovery, mechanical recovery, the dislocation model for the interpretation of the low actual shearing strengths of metals: this is only a very incomplete list of the epoch-making achievements that we owe to him. Polanyi's change of field was a great loss for metallurgy. A number of items of scientific and technically useful knowledge would probably have been attained more quickly with his collaboration than was possible without his help.

It should also be mentioned that the first years of Polanyi's activity in the field of metallurgical research came at a time of extreme economic depression. Just as he was for his collaborators the paradigm of the scientist constantly seeking for fundamental explanation, so, along with his charming wife, he also taught them to bear with good humour or even to overlook altogether the difficulties and limitations of the time.

NOTES

1. R.O. HERZOG and W. JANCKE, *Z.f. Phys.*, *3*, 1920, p. 196; *Ber. d. D. chem. Ges.*, *53*, 1920, p. 2162; P. SCHERRER in Zsigmondy, *Kolloidchemie* 3. Aufl.
2. M. POLANYI, *Naturw.*, *9*, 1921, pp. 288, 337; M. POLANYI and K. WEISSENBERG, *Z. f. Phys.*, *9*, 1922, p. 123; *Z. f. Phys.*, *10*, 1922, p. 44.
3. K. BECKER, R. O. HERZOG, W. JANCKE, and M. POLANYI, *Z. f. Phys.*, *5*, 1921, p. 61.
4. M. ETTISCH, M. POLANYI, and K. WEISSENBERG, *Z. f. Phys.*, *7*, 1921, p. 181.
5. E. V. GOMPERZ, *Z. f. Phys.*, *8*, 1922, p. 184.
6. H. MARK, M. POLANYI, and E. SCHMID, *Z. f. Phys.*, *12*, 1922, pp. 58, 78, 111.
7. H. MARK and M. POLANYI, *Z. f. Phys.*, *18*, 1923, p. 75.
8. M. POLANYI and E. SCHMID, *Z. f. Phys.*, *32*, 1925, p. 684.
9. M. POLANYI, *Z. techn. Phys.*, *6*, 1925, p. 121.
10. G. MASING and M. POLANYI, *Z. f. Phys.*, *28*, 1924, p. 169.
11. M. POLANYI and E. SCHMID, *Z. f. Phys.*, *16*, 1923, p. 336.
12. M. POLANYI, *Z. f. angew. Math. u. Mech.*, *5*, 1925, p. 125; *Z. f. Krist.*, *61*, 1925, p. 49; *Naturw.*, *16*, 1928, p. 285.
13. W. MEISSNER, M. POLANYI, and E. SCHMID, *Z. f. Phys.*, *66*, 1930, p. 477.
14. M. POLANYI and G. SACHS, *Z. f. Phys.*, *33*, 1925, p. 692.
15. P. BECK and M. POLANYI, *Naturw.*, *19*, 1931, p. 505.
16. W. EWALD and M. POLANYI, *Z. f. Phys.*, *31*, 1925, p. 139.
17. W. EWALD and M. POLANYI, *Z. f. Phys.*, *28*, 1924, p. 29; *Z. f. Phys.*, *31*, 1925, p. 746.
18. M. POLANYI, *Z. f. Phys.*, *89*, 1934, p. 660.
19. G. T. TAYLOR, *Proc. Roy. Soc.*, *A145*, 1934, p. 362.
20. J. M. BURGERS, *Proc. Kon. Ned. Akad. Wet.*, *42* 1938, pp. 293, 378; N. F. MOTT, and F. R. N. NABARRO, 'Rep. on Strength of Solids', *Phys. Soc.*, *1*, 1948, London; cf. also A. H. COTTRELL, *Dislocations and Plastic Flow in Crystals*, Clarendon Press, Oxford, 1956.
21. G. MASING and M. POLANYI, *Ergebn. exakt. Naturw.*, *2*, 1923, p. 177.

4

Rates of Reaction

HENRY EYRING

Introduction

SINCE nothing observable happens at complete equilibrium all of our knowledge of equilibrium systems is an extrapolation of observations on systems at least slightly displaced from equilibrium. However, if we measure the equilibrium compositions in some chemical system by absorption spectra, using low intensity radiation, the displacement from equilibrium can be made negligible. On the other hand, in chemical reactions we frequently deal with systems in which a forward reaction may be far from balanced by the converse, back reaction. In spite of this imbalance the usual preferred procedure for estimating non-equilibrium reaction rates is to extrapolate the well understood equilibrium value of rates of reaction to the incompletely understood non-equilibrium situation. The best methods of making these extrapolations are the chief concern of theoretical kineticists today and we shall consider some of the results and some of the outstanding problems.

Atoms

According to classical electrodynamics an electron spiralling toward a nucleus emits radiation and should finally come to rest in contact with the nucleus at a negatively infinite potential. This does not happen. The discrete energy levels required by the observed discrete atomic spectra led Bohr, in 1913, to his famous quantization rule. According to this rule an electron is restricted to stationary states for which its angular momentum is an integral multiple of $\frac{h}{2\pi}$. This restriction together with the usual laws of mechanics explain the hydrogen spectrum where h is taken equal to Plank's quantum of action. Finally wave mechanics retains Bohr's stationary states and his quantization rule but the lowest state turns out to have zero angular momentum with the hydrogen atom dive-bombing the nucleus endlessly. The energy of such an electron is the sum of two terms. The potential energy remains low if the electron restricts itself to the immediate neighbourhood of the nucleus; on the other hand if the electron is restricted to a small volume its kinetic energy rises excessively. The stationary states are those compromises which give stationary values for the sum of the two energies. The fact that the kinetic energy goes up with the confinement is a kind of electronic claustrophobia. Chemical bonding brings about a new adjustment between kinetic and potential energy by allowing the two bonding electrons to circulate about both nuclei.

Bonds

In the bonding of two nuclei by electrons the virial theorem requires that the potential energy decrease twice as fast as the kinetic energy increases. The extra manœuvre space provided by a second nucleus allows the bonding electron to visit the low potential internuclear space thus lowering the potential energy without an excessive rise in the kinetic energy. In accord with the Pauli exclusion principle two electrons inhabiting any orbital must have unlike spins. This excludes a third electron from entering any bonding orbital since of necessity the new-comer must have its spin parallel to one or the other incumbent. The third electron has no choice but to increase its energy and rise into a higher unfilled orbital.

Activation Energies

Because of the exclusion principle any atom or molecule bumping into a fully bonded molecule will meet resistance, since the oncoming electrons must be promoted to higher energy orbitals as the populated orbitals commence to overlap. If a collision is sufficiently violent the overlapping will proceed to a point of no return, the transition state. This point is reached when the approaching nuclei overlap orbitals as much as the initially bonded nuclei do. If the electrons pass over smoothly from the old to the new way of bonding, the process is said to be *adiabatic*, otherwise *diabatic*. Many reactions proceed adiabatically. When a reaction involves a change of multiplicity it can only succeed if there is also an electronic reorganization, as the molecules approach each other. Thus if triplet O_2 combines with singlet H_2 to make singlet H_2O_2 we have a diabatic reaction. The reaction $N_2O(\Sigma^1) \rightarrow N_2(\Sigma^1) + O(^3P)$ is a diabatic process where electronic reorganization adds a sizable improbability factor of the order of 10^{-3}.

Potential Surfaces in Configuration Space

Fritz London,[1] in the late 'twenties, proposed an approximate quantum mechanical formula for calculating the potential energy of three and four atom adiabatic reactions. Polanyi and the writer[2] developed this into the 'semi-empirical method' for calculating potential surfaces. This method was used to construct potential energy surfaces for a wide variety of reactions. With the development of these surfaces in configuration space the stage was set for applying statistical mechanics in an unambiguous fashion to reaction kinetics. The potential energy surface for a reaction involves as many dimensions as are required to fix a configuration plus one more for the energy. For a non-linear system of three atoms this means a four-dimensional space, for six atoms a seven-dimensional space and in general $3n-5$ dimensions are required unless the equilibrium configuration is linear in which case one more dimension is required to fix a configuration. Here n is the number of atoms. Besides the lowest surface on which most reactions take place there are an infinity of higher excited surfaces. For each change in one or more of the $3n-6$ quantum numbers, one for each co-ordinate, there exists a new surface. The excited surfaces are important in photochemical reactions and in reactions involving ions as observed, for example, in the mass spectrograph.

In general characteristics the potential surfaces are much like a landscape with stable molecules corresponding to valleys. These valleys are connected by passes in the region near the origin where the valleys approach each other. The saddle point along the pass connecting two valleys is the point of no return and is called the transition or activated state. The molecular complex corresponding to the transition or activated state is called the transition or activated complex. Distance along a valley and up through the pass into a neighbouring valley is said to be distance along the reaction co-ordinate. On the origin side of the reaction co-ordinate the energy (height) increases without limit as the distance between any pair of particles continues to shorten below the equilibrium distance. On the side of the reaction co-ordinate away from the origin the surface rises to a plateau corresponding to dissociation into atoms.

Absolute Reaction Rates

At potential minima, corresponding to stable compounds, and at saddle points, corresponding to the activated complex, it is possible to solve for the normal co-ordinates using the well-known theory of small vibrations. The resulting distances, frequencies, and energies provide all the quantities which go into the partition functions and so permit the calculation of equilibria and of all thermodynamic properties. For the activated complex the imaginary frequency corresponding to motion along the reaction co-ordinate can be treated as a translation provided an appropriate, usually small, factor for barrier leakage is introduced. A straightforward application of statistical mechanics then gives for the specific reaction rate constant the value

$$k' = \kappa \frac{kT}{h} e - \frac{\Delta G^{\neq}}{RT} \tag{1}$$

Here the free energy, $\Delta G^{\neq}$, is the standard free energy change per mole, i.e. the work that must be done to change the reaction products into the activated complex. In calculating $\Delta G^{\neq}$ the partition function for the activated complex reaction co-ordinate is taken equal to unity; $\dfrac{kT}{h}$ is then the frequency of reaction for the activated complex. At equilibrium κ is unity. Under non-equilibrium conditions κ differs from unity and corrects for barrier leakage, for quantum mechanical reflection and for failure to maintain equilibrium between reactants and activated complex. If it were not that κ is frequently near unity equation (1) would not be very useful. Wigner and Pelzer[3] were first to use the Eyring-Polanyi potential surface to calculate the rate of the reaction

$$H_2 \text{ (para)} + H \rightarrow H_2 \text{ (ortho)} + H. \tag{2}$$

This was followed by a general formulation by Eyring and by Evans and Polanyi,[4] who paid especial attention to pressure effects.

Transmission Coefficients

When two hydrogen atoms bump into each other in the absence of a third body they almost certainly separate again, since the oscillator is non-polar and so incapable of radiating dipole radiation. κ for this association reaction is consequently extremely small, as is necessarily true also for the converse unimolecular decomposition of H_2. The dissociation of a diatomic molecule or the association of two atoms is thus very sensitive to collision with a third body in the form of gas molecules or surfaces. The decomposition of a polyatomic molecule on the other hand ceases to increase with pressure whenever collisions are frequent enough to maintain the equilibrium concentration of energy-rich molecules against the drain of the unimolecular decomposition. Enough pressure to provide pressure insensitiveness is no guarantee that every pair of colliding radicals will stick together. To stick, the molecules must be properly oriented and be otherwise adjusted to each other's motions. The evaporation or sublimation of molecules from a surface is an especially instructive unimolecular decomposition as is the converse association. The accommodation or sticking coefficient of a molecule colliding with its liquid surface is observed to be the ratio of the value of the rotational partition function in the liquid to the value in the gas.[5] This is understandable if sticking to a surface is nearly adiabatic with respect to rotational energy, so that only those molecules whose gaseous rotation can go adiabatically into an appropriate rotational or rocking motion in the liquid can stick. Stated another way, this means that in a unimolecular decomposition we should use for the departing radicals or molecules in the activated state the rotational partition functions they had in the initial state. This is in spite of the fact that in the activated state the potential energy restriction against rotation of an evaporating molecule no longer exists. κ in equation (1) for vaporization is thus

Rates of Reaction

$\kappa = f_e/f_g$ where f_e and f_g are the rotational partition functions for liquid and gas respectively. Earl Mortensen and the author were thus led to the conception of a pool at equilibrium in front of the rate determining barrier. The reacting system passes over the barrier adiabatically with respect to the rotational states of the separating radicals and to some extent with respect to internal vibrations and the overall rotation of the activated complex. Thus in unimolecular decompositions the transmission coefficient, κ, can be quite different from unity and, at present, an active phase of reaction kinetics is the effort to specify κ ever more closely.

Simultaneous Chemical Reactions

Long ago Christiansen, Herzfeld and Polanyi[6] showed that the hydrogen and bromine reaction studied by Bodenstein and Lind[7] must be explained as a chain of reactions in this case involving bromine and hydrogen atoms. Since that time many systems have been found to proceed by such reaction chains. For each species with a concentration C_i an equation

$$\frac{dc_i}{dt} = \sum_j k_{ji} c_j - \sum_j k_{ij} c_i \tag{3}$$

can be written where a particular k involves as factors concentrations of other species if the corresponding process is other than unimolecular. The set of equations (3) is usually solved by setting $\frac{dc_i}{dt} = 0$ for all unstable intermediates. The k's in equation (3) each involve also as one factor the appropriate elementary specific rate constant discussed in connection with equation (1).

Exact Reaction Rate Theory

It is natural to ask if it is possible to get away from the non-equilibrium complications that we have lumped together in the transmission coefficients, κ, by letting the species in the set of equations (3) be single quantum states. According to quantum mechanics this is possible. However, when one sets about looking for experimental proof that the k's do not deviate from their equilibrium value the evidence is much more scanty than one could wish. Consider the following simple test. If i is a pure state with transitions to the pure states j and k then the corresponding spectral line intensities $I_{i \to j}$ and $I_{i \to k}$ are $I_{i \to j} = C_i k_{ij}$ and $I_{i \to k} = C_i k_{ik}$. The ratio of intensity R for the two lines is

$$R = \frac{I_{i \to j}}{I_{i \to k}} = \frac{k_{ij}}{k_{ik}} \tag{4}$$

The question is thus whether the ratio, R, of intensities of the two lines is independent of the manner of recruiting the population in i. Constancy of R is understandable if the population in state i completely forgets where it is recruited from, i.e. if each specimen of the population attains complete randomness in phase. Spectral lines usually have finite widths with different adsorption coefficients for centre and edge. Thus the k's vary for the different parts of a line so that it appears problematical whether the k for an equilibrium system applies under non-equilibrium conditions. If k_{ij} is not independent of the manner of populating state i the sets of equation (3) become essentially impossible to solve exactly since every k_{ij} then depends on the concentrations of all the other species. Even in such an unhappy contingency using equilibrium values for k_{ij} should lead to a very good approximation.

Reaction Rates in Living Things

On earth, living things are made of l-amino acids and d-sugars. If our world could be suddenly changed by reflecting it in a mirror, it would contain d-amino acids and l-sugars, but in other respects it would carry on as before. The history of this world and that of its mirror image would be faithful copies of each other. Amino acids are fashioned into protein by ribonucleic acids aided by enzymes which are themselves proteins. Desoxyribonucleic acid, which constitutes the genetic material present in the chromosomes, differs from ribonucleic acid chiefly in containing ribose in the reduced instead of the unreduced form. The outstanding characteristic of the protein enzymes, the ribonucleic acids and the desoxy-ribonucleic acids is that together they fashion other molecules in their own image and multiply endlessly. Only the l-amino acids and the d-sugars fit against the templets made from molecules like themselves. In the stream of living things the occasional protein molecule with mixed d and l-amino acids lacks survival value and so fails to be reproduced competitively. On the other hand, the endless variety of living things indicates that all sorts of mutations have subsequently emerged from the original successful ribonucleic acid and protein molecules which won out in competition with their mirror images by first perfecting themselves to the point where they could reproduce by replication. This race of our l-amino world against the equally likely d-amino world must have been won about 1500 million years ago.

We turn next to the exciting problem of investigating enzymes in living systems. By studying luminescence it is possible to obtain a fascinating insight into the working of enzymes in general and of the oxidative enzymes in particular. The emission of light by all creatures studied such as the firefly, the glow-worm, cypridina and luminous bacteria depends on oxygen molecules taking electrons from a metabolite called luciferin with the aid of an enzyme called luciferase. Ninety-nine times out of a hundred the outer, more loosely bound, electrons are removed but the hundredth time one of the outer electrons is left behind. After gyrating briefly this electron drops into the hole left by removal of the more tightly held electron, and in doing so emits a quantum of light. An oxygen molecule is unable to react with luciferin except as it is aided by the enzyme luciferase; so by studying the effect of different agents on light emission we are in effect studying their influence on the enzyme. Luminous bacteria emit almost no light at 0° C., but as the temperature is raised the light intensity rises exponentially, as Arrhenius predicted, passes through a maximum around 17° C., and then drops to near zero around blood temperature. The drop in light intensity, with rise in temperature, is due to the denaturation of the luciferase, which unfolds, exposing its hydrophobic groups. In its unfolded state it can no longer function as an enzyme. By exposing its hydrophobic groups, the enzyme repels the water molecules and the systems as a whole expands. The LeChatelier principle then tells us that the application of hydrostatic pressure will reverse this denaturation and increase the light emitted. This is found to be true. However, in the low-pressure range, pressure decreases the light emitted. This proves that the activated state is more voluminous than the folded state of the enzyme, indicating that the activated state is partially unfolded. Now molecules which are made up of a hydrophobic part and a hydrophyllic part, like the alcohols, ketones, ethers, etc., should promote unfolding of the enzyme by juxtaposing their hydrophobic groups with the exposed hydrophobic groups of the denatured enzyme, thus promoting denaturation and thereby stopping luminescence. This in fact happens. Three alcohol molecules are required to inactivate one luciferase molecule. The cure for inactivation is equally obvious. The application of pressure should reverse the denaturation and turn the light on. It does. Johnson and Flagler, acting on the theory that the oxidative enzymes responsible for consciousness are inactivated by alcohol, just as are the oxidative enzymes responsible for luminescence, placed tadpoles or salamanders in a pressure vessel and added about $1\frac{1}{2}$ per cent of alcohol to the water which made them drunk. Then by a stroke or two of the pump they raised the pressure to around 200 atmospheres, when the tadpoles started swimming again. With release of pressure they immediately returned to a drunken stupor. This could be repeated a number of times before serious damage was done. Sulfanilamide and para-amino benzoic acid are likewise effective in inhibiting luminescence. Much smaller doses of these are required than of the alcohols. These agents act by being preferentially adsorbed on the active sites of the enzyme. Pressure is without appreciable effect in reversing this inactivation since no unfolding of the enzyme is involved. Irreversible denaturation of enzymes as well as many other phenomena have been treated elsewhere from this kinetic point of view.[8]

Other Pressure and Temperature Effects on Reaction—Diamond Formation and Mountain Building

Diamonds at absolute zero and one atmosphere pressure are less stable than graphite. If the pressure is raised to about 15,000 atmospheres diamonds become the stable phase. At absolute zero the rate of conversion is infinitesimal. As the temperature is raised to speed up graphite conversion to diamond the pressure must likewise be raised to keep diamond the stable phase. Bundy, Hall, Strong, and Wentorf at General Electric have developed the diamond synthesis to a routine affair. A thin film of metal around the diamond is required as a catalyst to break the bonds in graphite by dissolving carbon atoms so that they can rapidly re-form in the diamond lattice. In the absence of the metal catalyst the activated complex for the diamond to graphite transition is so voluminous that pressures insufficient to make diamond thermodynamically stable still suppress graphitization by making diamond kinetically stable.[9]

Mountain building seems to involve analogous phase transitions. For a mountain to exist in approximate isostatic balance, it must have deep roots of some light mineral such as basalt, whose density is 2·4 to 2·5, floating in some heavy mineral like eclogite, of density 2·8 to 2·9. As in the diamond-graphite equilibrium, low pressures and high temperatures presumably act to stabilize the light phase. Thus the less dense phase is to be expected to extend to greater depths, forming highlands wherever some source of heat such as uranium maintains an unusually high temperature. It is therefore not surprising that the aluminium oxide silicates rocks, with their affinity for radioactive materials, should form the continents, whereas the ocean basins should be less rich in radioactive material. The Mohorovicic transition between the light and dense phase is near the surface over the ocean and deeper, ordinarily, over the continents, as is to be expected. However, this too simple explanation is complicated by chemical variations modifying the mineralogical phase transitions and by convection currents underneath the surface of the earth which depress some areas and elevate others creating notable gravitational anomalies to complicate the simple isostatic picture. Kennedy has recently presented interesting evidence for the role of phase transitions in mountain building.[10] This whole matter becomes of great interest from the reaction rate point of view now that high temperature-high pressure techniques enable us to explore such problems experimentally.

Transport Properties in Condensed Phases

In both solids and liquids dislocations, vacancies and other imperfections allow localized displacements in the relaxation of stress. The more plentiful the appropriate type of imperfection the more rapid relaxation occurs. The expansion accompanying melting is about 12 per cent for normal liquids like argon, 3 per cent for metals and 22 per cent for sodium chloride. Water contracts because a change in packing superposes a 20 per cent contraction on the normal 10 per cent expansion accompanying melting. Fluidity in liquids is proportional to expansion, as Batschinski pointed out. This is presumably because the number of dislocations speeding up relaxation is proportional to the expansion. If the distance between atoms on the two sides of a shear plane is λ_1 and each displacement of a patch of area $\lambda_2\lambda_3$ is through a distance λ, the rate of shear is

$$\dot{S} = \frac{\lambda 2k'}{\lambda_1} \sinh \frac{f\lambda_2\lambda_3\lambda}{2kT} = \beta^{-1} \sinh f\alpha \tag{5}$$

where k' is the rate of relaxation of the patch under zero shear stress and f is the shear stress. If equation (5) is solved for f we obtain

$$f = \frac{1}{\alpha} \sinh^{-1}\beta\dot{S} \tag{6}$$

and the viscosity is

$$\eta = \frac{f}{\dot{S}} = f\beta \, (\sinh \alpha f)^{-1} \tag{7}$$

If $\alpha f \gg 1$ we obtain

$$\eta = \beta/\alpha \tag{8}$$

which gives a good account of viscosity of simple systems.[11] Now if a shear plane has different types of relaxation mechanisms where the ith type occupies a fraction of the area x_i we have for the viscosity

$$\eta = \sum_i x_i f_i/\dot{s} = \sum_i \frac{x_i \beta_i}{\alpha_i} \frac{\sinh^{-1}(\beta\dot{S})}{\beta\dot{S}} \tag{9}$$

Equation (9) gives an excellent account of the viscosity of all sorts of systems. Now, as Einstein showed, the viscosity and the diffusion, D, are related by the equation

$$\eta D = \frac{kT}{6\pi r} \frac{d\ln a}{d\ln c} \tag{10}$$

provided the diffusing ion is large compared with the viscous solvent. For self diffusion Sutherland corrected equation (10) to

$$\eta D = \frac{kT}{4\pi r} \frac{d\ln a}{d\ln c} \tag{11}$$

Here a, c and r are the activity, concentration and radius of the diffusing substance. Both equations (10) and (11) agree quite well with experiment in many cases. Relaxation theory yields the analogous equation for self diffusion

$$\eta D = \frac{kT}{\zeta\lambda} \frac{d\ln a}{d\ln c} \tag{12}$$

Here λ is a mean lattice distance and ζ is six for close packed liquids, since this is the number of near neighbours lying in the same plane with which a molecule can articulate separately as it diffuses normal to the plane. Equation (12) agrees with equation (11) and correlates very well with a great deal of data. We shall not attempt to carry this brief survey of transport properties further, but there is a tremendous literature for the interested reader.

Reactions Involving Ions—Mass Spectrometry, Strain Electrometry and Smelting

When electrons are shot into molecules with more energy than is required to cause ionization positive ions are formed. In accord with the Franck-Condon principle, the ionization occurs prior to any appreciable change in the internuclear distances. Since the equilibrium internuclear distances are different for a molecule than for its positive ion, it follows that the ions are created with excess potential energy. The ion first formed by the loss of only an electron from the original molecule is called the parent ion. The excess energy, except as it causes instantaneous decomposition, distributes itself more or less randomly throughout the ion. Whenever sufficient energy accumulates in the bond or bonds holding two parts of the ion together to break it in two, a daughter positive ion and a molecule or radical are formed by the fission. The mass spectrometer collects the positive ions that appear approximately 10^{-6} seconds after electron bombardment. Ions which decompose after falling through the accelerating field but before traversing the magnetic field are recognizable because of fractional values of $\frac{e}{m}$.

The minimum energy which the bombarding electrons must have in order for a particular ion to appear is called the appearance potential. The difference between the appearance potential of daughter and parent ion gives the activation energy for the daughter. With the known activation energies and reasonable estimates of vibrational frequencies and moments of inertia one can apply absolute rate theory to calculate the relative rates of fission into the different possible ions appearing in the mass spectrometer providing one can measure, calculate or guess successfully the distribution of energy in the parent ions. The initial paper outlining this theory[12] has started investigations which will not be developed further here. In mass spectrometry we have an especially powerful tool to evaluate theories of unimolecular decomposition[13] as well as theories involving collisions with ions.

Reactions at interfaces frequently transport ions. Any imbalance between the rates at which positive and negative current is transported at the existing potential initiates an electrical transient in which the potential grows until it balances the rates for positive and negative charge transport. This is the source of the ·1 volt potential across the membranes of all living cells. Analogously, when a weight of about a kilogram is suddenly added to a 20-mil aluminium wire immersed in water, a transient potential of about one volt builds up against a similarly situated unstrained wire in about ·1 second having a half life of about half a second, depending on circumstances. The strained wire is negatively charged. By changing the nature of the wire, its treatment, the solution, the surrounding atmosphere, the stirring, the temperature, etc., one gets almost every conceivable modification in the electrical transient. This field of strain electrometry is another fertile field of application of absolute reaction rate theory.[14]

At the interface between molten iron containing carbon, sulphur and other impurities and the slag above one has oxide ions from the slag combining with carbon in the iron to give carbon monoxide. This leaves the electrons as a charge on the iron. These electrons are carried back into the slag as sulphide ions and are neutralized by positive ions in the slag, such as quadrivalent silicon and divalent iron, diffusing over and becoming part of the metal, thus using up the excess electrons. This electrochemical nature of smelting requires that all the electrical currents balance as was demonstrated by King and Ramachandran.[15] Using absolute reaction rate theory X. de Hemptinne, T. Ree and the author were able to use the observed rate of evolution of CO together with the various concentrations to predict the course of reaction for all of the electrode reactions.

In this quick Cook's tour of the fields of chemistry which are of current interest to the author, one cannot fail to note that Polanyi, although trained as a physician, played a leading role in chemical kinetics during the more than thirty years it occupied his attention. As one who profited by his sage and always kindly advice, one can only say that his more recent eminence in new fields was to be expected, and the end is not yet.

NOTES

1. F. LONDON, 'Probleme der Modernen Physik', *Z. physik, Sommerfeld Festband*, 1928, p. 104.
2. H. EYRING and M. POLANYI, *Z. physik. chem. B.*, *12*, 1931, p. 279.
3. H. PELZER and E. WIGNER, *Z. physik. chem. B 15*, 1932, p. 445.
4. H. EYRING, *J. Chem. Phys. 3*, 1935, p. 107
   M. G. EVANS and M. POLANYI, *Trans. Faraday Soc. 31*, 1935, p. 875.
5. G. WYLLIE, *Proc. Roy. Soc. A 197*, 1949, p. 383.
6. J. A. CHRISTIANSEN, *Kgl. Danske Videnskat. Selskab., Matt.-fys. Med. 1*, 1919, p. 14.
   K. F. HERZFELD, *Ann. Physik 59*, 1919, p. 635.
   M. POLANYI, *Z. Elektrochem. 26*, 1920, p. 50.
7. M. BODENSTEIN and S. C. LIND, *Z. physik. chem. 57*, 1907, p.168.
8. F. H. JOHNSON, H. EYRING, and M. J. POLISSAR, *The Kinetic Basis of Molecular Biology*, New York, John Wiley and Sons, Inc., 1954.
9. P. W. BRIDGMAN. *Proc. Amer. Acad. Arts and Sci.*, 76, 1945, p. 1.
   H. EYRING, and F. WM. CAGLE, JR. *Zeit. f. Elektrochemie*, 56, 1952, p. 480.
10. G. C. KENNEDY, *American Scientist*, 47, 1959, p. 491.
11. T. REE, and H. EYRING, *Rheology*, Edited by Eirich, New York, Academic Press, 1958, ch. 3.
12. H. M. ROSENSTOCK, M. B. WALLENSTEIN, A. D. WAHRHAFTIG, and H. EYRING, *Proc. Nat'l. Acad. Sci.*, 38, 1952, p. 667.
13. N. B. SLATER, *Theory of Unimolecular Reactions*, Ithaca, N.Y., Cornell Univ. Press, 1959.
14. A. G. FUNK, J. C. GIDDINGS, C. J. CHRISTENSEN, and H. EYRING, *Proc. Natl. Acad. of Sci.*, 43, 1957, p. 421.
15. T. B. KING and S. RAMACHANDRAN, *The Physical Chemistry of Steelmaking*, pp. 125–46, New York, John Wiley and Sons; and London, Chapman & Hall, 1958.

5

The Size and Shape of Molecules, as a Factor in their Biological Activity*

ERNST D. BERGMANN

IN a number of classical papers, M. Polanyi has introduced into the theory of the kinetics of chemical reactions an approach which might justifiedly be called microgeometrical. Describing the dependence of the reaction velocity of various organic halides with sodium atoms in the vapour phase as a function of structure, he postulated that the sodium atom must approach the carbon-halogen bond, for electrostatic reasons, from the side of the halogen atoms, giving a transition state of the type C—Hal—Na, before forming sodium halide, NaHal. Only this picture of the reaction mechanism explained that the velocity determining factor is the strength of the carbon halogen bond, and not the structure of the organic moiety of the molecule.[1] The application of this principle has opened a new era in reaction kinetics; it has made understandable at least some cases of the complex phenomenon known as Walden inversion[2,3], *viz.* those in which an organic halide reacts with a negative or positive ion. In the former case, the ion—again for electrostatic reasons—must approach the carbon halogen bond from the side of the carbon atom

$$X^- + C - Hal \longrightarrow X - C + Hal^-;$$

the velocity will depend on the extent to which the other substituents of the carbon atom shield the reacting bond, and the reaction will lead to inversion of the configuration. In the latter case, the ion will approach the *C-Hal* bond from the side of the halogen atom

$$C - Hal + Y^+ \longrightarrow C^+ + HalY,$$

the velocity will—again—depend only on the strength of the bond, and the reaction will *not* be accompanied by inversion.

Polanyi's microgeometrical approach to the explanation of these

* The writer of these lines has been fortunate in being permitted to work with Professor Polanyi in the early 'thirties and has always remained conscious of the inspiration that emanated from this collaboration. This modest contribution, therefore, expresses not only admiration for a great scientist, but also deep personal gratitude.

reactions has been foreshadowed by some qualitative theories current in classical organic chemistry. The rendering of the specificity of enzymes for their substrates by the picture of key and lock is a case in point; however, this picture has only recently been developed into something more than an indistinct drawing. The same is true for Paul Ehrlich's conception of chemotherapeutical research as the search for a 'magic bullet' that will hit the infecting parasite or bacterium, without damaging the host cell.

It may be interesting to show how the microgeometrical approach has helped to develop these classical theories and to make them more specific, more quantitative and, therefore, more useful.

Ehrlich's theory would be applicable if one were to find a compound, a metabolite, which is essential for the infecting organism, but not for the host, and interfere with either its synthesis or utilization. That this is possible, has first been shown by Woods.[4] He explained the chemotherapeutical activity of p-aminobenzenesulphonamide as interference with the utilization of p-aminobenzoic acid which is an essential growth factor for some pathogenic bacteria and has to be supplied to them through the medium in which they grow; this amino-acid is, e.g. necessary as a building element for the synthesis of folic acid by the bacteria, whilst the human body does not require an external supply of it. The bacterium cannot distinguish between the metabolite p-aminobenzoic acid and the sulphonamide *because of their geometrical similarity*; it tries to use the latter instead of the former, but the products of this utilization have no longer the biochemical properties of the 'normal' analogs based on p-aminobenzoic acid and cannot exercise their functions: the sulphonamide is an *anti*metabolite. This theory is supported by the observations that bacteria which do not require p-aminobenzoic acid are not affected by p-aminobenzenesulphonamide, and that large doses of the former relieve the lethal effect of the latter even in such bacteria as do require the metabolite. It seems obvious that the relationship between the two compounds is a competitive one; the mechanism is governed by the law of mass action.

This theory has led to the discovery of a number of antimetabolites, which are all characterized by their close similarity in shape and size to certain metabolites, and are thus able to interfere with them. This may be illustrated by the following structural formulae:

$$
\begin{array}{cc}
\text{Metabolite} & \text{Antimetabolite} \\[6pt]
\underline{CH_3S\text{-}CH_2CH_2}\text{-}\underset{NH_2}{CH}\text{-}COOH & \underline{CH_3\,CH_2S\text{-}CH_2}\text{-}CH_2\text{-}\underset{NH_2}{CH}\text{-}COOH \\[6pt]
\text{Methionine} & \text{Ethionine}^5 \\[6pt]
 & \underset{\underline{O}}{\overset{\text{C}H_3\text{-}S\text{-}CH_2\text{-}CH_2\text{-}\underset{NH_2}{CH}\text{-}COOH}{}} \\[6pt]
 & \text{Methionine sulphoxide}^5
\end{array}
$$

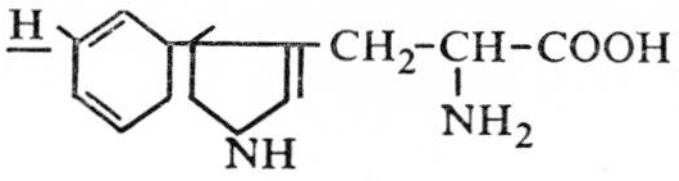

The Size and Shape of Molecules, as a Factor in their Biological Activity

Tryptophan

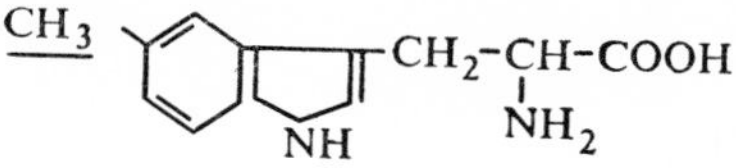

5-Methyltryptophan[6]

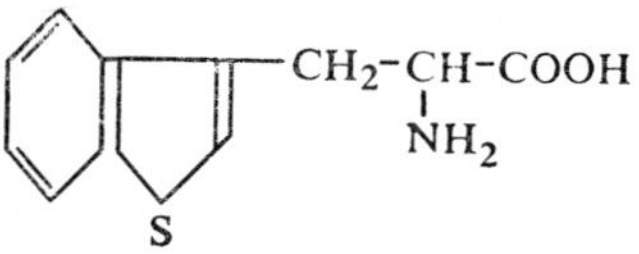

β-(3-Thionaphthenyl) - alanine[7]

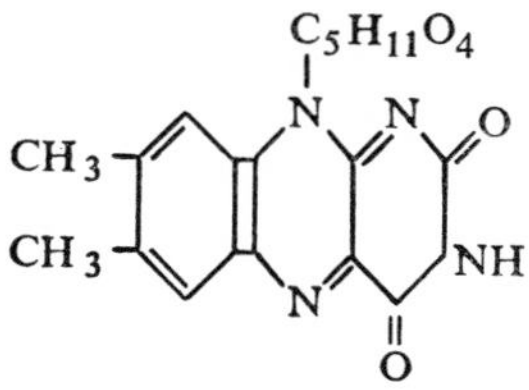

Riboflavin (vitamin B$_2$)

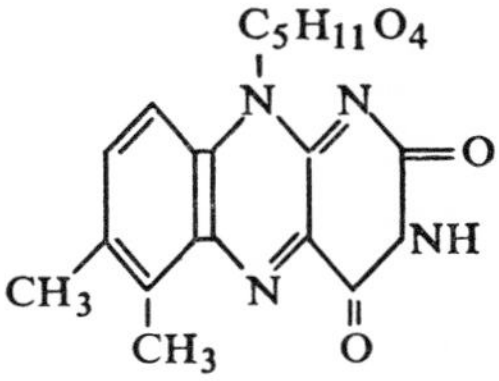

"Isoriboflavin"[8]

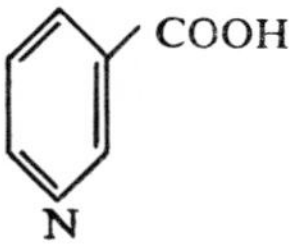

Nicotinic acid

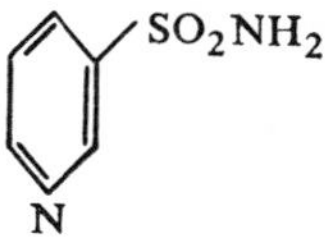

Pyridine-3-sulphonamide[9]

Clearly, the application of the antimetabolite principle is not restricted to bacterial metabolism; it may apply to every living cell. Indeed, one of the most promising approaches to cancer chemotherapy has been based on the need of cancer cells for pyrimidines and purines as constituents of their nuclei. Thus, azathymine, azacytosine, and azauracil must be considered as antimetabolites of the normal cell constituents thymine, cystosine, and uracil, 6-mercaptopurine and 8-azaguanine as antimetabolites of adenine and guanine. A more recent development, initiated by Woolley,[10] points to the usefulness of the antimetabolite theory and, therefore, of the microgeometrical approach in the treatment of mental disturbances. Experimental schizophrenia can be induced by the diethylamide of lysergic acid, a constituent of the ergot alkaloids. This may be due to the antimetabolite action of this compound on serotonine and adrenaline, essential constituents of the nervous system. In fact, lysergic acid can be considered as a modified adrenaline or serotonine.

[Structures: Adrenaline, Lysergic acid, Serotonine]

In confirmation of this hypothesis, some other substances have been synthesized which resemble lysergic acid in their structure sufficiently to simulate its activity, e.g. N-methyl-N-propyl-1,2,3,4-tetrahydro-2-naphthylamine.[11]

A very good example for the usefulness of the microgeometrical approach is the observation that a metabolite can be converted into an antimetabolite by replacement of a hydrogen atom by fluorine. In those cases at least to which this statement applies, the fairly large change in electrochemical character, in polarity, brought about by this substitution, has no immediate effect, and the effective radius of fluorine (1.35A) is sufficiently close to that of hydrogen (1.20A), so that the change appears not to affect perceptibly the geometrical shape of the molecule. (In fact, the fluorine atom is the smallest of all possible substituents and, therefore, in many aromatic compounds, hydrogen can be replaced isomorphogeneously by fluorine.[11a]). The living cell *incorporates* the fluorine compound instead of the parent substance, and sometimes the fluorinated product so formed participates in further steps of the metabolism, until the cell 'becomes aware' of the anomaly. In these cases, a 'lethal synthesis' has taken place, to use a phrase coined by Peters:[12] the antimetabolite itself does not affect adversely the function of the cell, but is catabolized to a compound that does. A case in point is fluoroacetic acid, which is transformed by the pathway of the Krebs cycle into fluorocitric acid; this acid stops the further functioning of the cell by inhibiting the cycle at the stage of the enzyme aconitase.[12] Several other cases are known in which fluorine compounds are incorporated into cell constituents instead of the halogen-free parent substances, e.g. *p*-fluorophenylalanine,[13] *o*-fluorophenylalanine,[14] and 5-fluorouracil (probably in form of its riboside);[15] and the fact that 5- and 6-fluorotrytophan[16] are substrates for the enzyme that 'activates' trytophan for its incorporation into cell protein, can very probably be interpreted analogously. Contrary to this behaviour, 5-methyltrytophan ('methyl radius' 1·77Å) is not activated by the above enzyme system,[16] and neither this isomer[17] nor 6-methyltrytophan[13] is incorporated by *E. coli*. Equally, the larger diameter of the chlorine atom (1·89Å) makes it impossible for *p*-chlorophenylalanine to be incorporated into the protein of this bacterium.[13]

To be sure, one has to distinguish a number of different possibilities, all of them compatible with the similar sizes of hydrogen and fluorine atoms. The fluorine compound can be an inhibitor as such; it can, e.g. combine with the enzyme which normally adsorbs and transforms the metabolite and, thus, prevent this 'normal' reaction. It can, secondly, be metabolized by this enzyme in the same manner as the normal metabolite and thus enter the pathway of the 'lethal synthesis' of an inhibitor. It can also be without any deleterious effect, which again may mean one of two things: the cell incorporates the fluorine compound without any effect on its normal functions, or it remains completely refractory to the fluorine compound. It may well be that all of these possibilities materialize in different cases, but not enough is known about the fate of the many fluorine compounds tested to be quite sure about this point. [17a]One might conclude that whenever the fluorinated analog behaves like the fluorine-free parent compound, the shape of the molecule governs the biological reaction in question; whenever the two substances differ, electrochemical factors play a decisive part. 3-Fluoro-4-aminobenzoic acid is a competitive antagonist of *p*-aminobenzoic acid, the 2-fluoro-derivative replaces the 'normal' acid as bacterial growth factor.[18]

The microgeometrical approach has recently been used in a number of other cases; it may well be that in fact these cases also are only examples of the principles enumerated above. Such a case is the hypothesis for the mode of action of the carcinogenic polycyclic hydrocarbons,[19] compounds that are highly active without containing any functional groups. The fact that only substances which are completely planar molecules of a certain overall size are carcinogenic, has made the conclusion most attractive that the carcinogenic molecule must 'fit' a certain surface, e.g. of an enzyme, and can thus block the metabolic processes in which this surface is involved.

A similar conclusion has been drawn in the attempt to elucidate the mode of action of the insecticide D.D.T. In correlating the contact toxicity of this compound and its analogs with their structure, the 'classical' approach, which divided the molecule into a lipoid-solubilizing part ('haptophor' in Paul Ehrlich's terminology) and a toxic part, seemed to be very artificial, considering the formula of the D.D.T. molecule. It has, therefore, been proposed that the geometrical structure of the molecule determines its activity, in particular the free rotability of the two benzene rings and the CCl_3 group.

E. D. Bergmann

$$\text{Cl-C}_6\text{H}_4\text{-CH(CCl}_3\text{)-C}_6\text{H}_4\text{-Cl}$$

Every substitution which adversely affects this feature, destroys the insecticidal activity.[20] It must be emphasized that this hypothesis is not an 'explanation' of the biological activity of the compound; it only relates this activity to a defined microgeometrical factor.

How, in fact, does the 'microgeometrical' approach lead to a better understanding of the biological activity of chemical compounds? One will have to assume—and we have done so more or less tacitly in the preceding paragraphs—that the geometrical structure of the active molecule corresponds to the structure of a biological receptor, e.g. an enzyme. We have no direct knowledge of the active centre of such proteins, especially those which are active without having any outstanding prosthetic groups, and we must assume that they contain amino acids in a given order (distance from each other) and geometrical arrangement which are responsible for the activity. Only indirectly, from the structure of the drug, can we gain an insight into the nature of the receptor.[21]

However, the fact that enzymes react differently to geometrically isomeric compounds and thus may respond to steric factors, has been known before, at least implicitly. Fumaric acid has been known[22] to be a stronger inhibitor of such enzymes as diastase and lipase than maleic acid, an observation which will easily be related to the well-established fact that the former acid is much better absorbed on solid surfaces than the latter.[23] Recently, the *cis*- and *trans*-isomers of N-substituted 2-aminocyclohexanols have been reported to act differently on acetylcholinesterase.[24]

An attempt to deduce the active site of the enzyme from the structure of the substrate has been made successfully in a number of cases, two of which will be discussed here by way of example.

Acetylcholinesterase splits acetylcholine according to the scheme

$$(CH_3)_3N^+-CH_2-CH_2-O-\underset{CH_3}{C}=O \longrightarrow (CH_3)_3N^+-CH_2-CH_2-OH + HO\underset{CH_3}{C}=O$$

If one makes the reasonable assumption that the substrate will have to form a complex or compound with the enzyme, it follows that the latter will most likely contain a negatively charged group or region which can bind the positively charged ammonium moiety of the substrate. This has, indeed, been proven by various means,[25] e.g. by comparative experiments with dimethyl-aminoethyl acetate

The Size and Shape of Molecules, as a Factor in their Biological Activity

$$(CH_3)_2N-CH_2-CH_2-O-\underset{\underset{CH_3}{|}}{C}=O$$

which is split rapidly only in acidic solution, i.e. in the cationic form

$$(CH_3)_2\overset{+}{N}H-CH_2-CH_2-O-\underset{\underset{CH_3}{|}}{C}=O$$

The distance of closest approach between the negative region of the enzyme and the ammonium nitrogen has been calculated from the Brönsted equation, relating the dissociation constants of ionic complexes to the magnitude and distance of the charges. The distance found was 6·3 Å; this would mean—if one recalls that the radius of the tetramethylammonium ion is 3·5 Å—that the radius of the negative region is 2·8 Å. We do not yet know the chemical nature of the carrier of this charge; it may be recalled that the radius of a negatively charged oxygen atom is 1·4 Å, that of a negatively charged sulphur atom 1·84 Å,[26] i.e. of a reasonably close order of magnitude. One might think of the (dissociated) free carboxyl group of glutamic or aspartic acid in the protein as seat of the negative charge.

Obviously, the enzyme must contain, in addition to the negative site, a grouping which is responsible for the hydrolytic reaction ('esteratic site'). This grouping will be expected to be of basic nature, so that it can attach and attack the carbon atom of the ester carbonyl, as in the base-catalyzed homogeneous saponification of esters. From the p_H dependence of the enzymatic reaction—the structure of acetylcholine is independent of p_H—conclusions can be drawn as to the nature of the enzyme protein, in particular of the esteratic site. It has, indeed, been possible to determine from such data the dissociation constant of the negatively charged and the esteratic sites; their p_H is 9·2 and 6·5, respectively. The latter value has been interpreted to mean that the esteratic site is an imidazole ring;[27] we know that acylimidazoles are very prone to hydrolytic fission. Finally, one may calculate the distance of the esteratic site from the negative region, from the distance of the carbonyl carbon atom and the quaternary ammonium ion. It is 2·5 Å. Thus in some measure the structure of the active part of acetylcholinesterase can be visualized; as the purpose of this essay is only to show the usefulness of the microgeometrical approach, we will refrain from an elaboration of some finer points, as, e.g. the meaning of the observation that the intermediary acylation of the enzyme by the acetyl group of acetylcholine (or the acyl radicals of similar compounds) leads, if the product is hydrolyzed and isolated, to an acylated serine (not histidine), or the nature of the forces acting between enzyme and substrate.

A second, recent example may be quoted instead, that of the active surface of yeast lactic dehydrogenase. The reasoning in this case has been indirect; it was based on the relationship between the structure and activity of inhibitors of the enzyme.[29] As all substrates and inhibitors of the enzyme are anions, its active surface must contain a positively charged point of attachment. Secondly, the logarithm of the ratio [velocity without inhibitor/velocity in presence of inhibitor] as a function of the inhibitor concentration is a straight line; from its slope one can deduce that one molecule of substrate (or inhibitor) reacts with one 'molecule' of the enzyme. Thirdly, the charge e on the protein surface, divided by the radius r of closest approach, was found to be $\frac{1}{11 \cdot 1}$ (e expressed in charge units). The four amino acids in peptide structures which carry positively charged units (at p_H 7·5) in the side-chain, are histidine, lysine, ornithine, and arginine. As the inhibition of the enzyme by p-toluenesulphonic acid was p_H independent in the range between 5·5 and 9·0, histidine was ruled out; as the enzyme was not inhibited by typical amino group reagents, lusine and ornithine must also be excluded. Thus, the guanidino-group of arginine reveals itself as the carrier of the positive charge. Indeed, for this group, in which the charge is distributed between two nitrogen atoms (see formula), $e = \frac{1}{11 \cdot 6}$ per nitrogen (and one can assume that one of the nitrogen atoms is neutralized by a neighbouring ionized carbonyl group). Taking into account that the prosthetic group of the enzyme is a flavocytochrome, one can formulate the enzyme-lactic acid complex as follows, making clear the transition to the (ion of) pyruvic acid:

Contrary to the previously discussed example, the substrate is here bound to the enzyme by a *three*-point attachment. It can be shown that for any enzyme which discriminates between *optical antipodes* of the substrate, i.e. acts asymmetrically, at least three points of attachment between enzyme and substrate are necessary. This statement[30] can be reversed: an enzymatic reaction which passes through a state of three-point attachment takes place asymmetrically, even if the substrate (in its isolated form) shows no molecular asymmetry.[31] This is the case for citric acid, which is synthesized from oxaloacetic acid and transformed in the Krebs cycle to α-ketoglutaric acid.

$$O=C\text{-COOH} \quad CH_3COOH \longrightarrow HO\text{-}C\text{-COOH} \longrightarrow HOOC\overset{*}{\text{-}}CO\text{-}CH_2\text{-}CH_2\text{-}COOH$$
$$\underset{CH_2\overset{*}{C}OOH}{} \qquad \underset{CH_2\overset{*}{C}OOH}{}$$
$$\text{citric acid}$$

The Size and Shape of Molecules, as a Factor in their Biological Activity

If the carbon atom of the γ-carboxyl of oxaloacetic acid was labelled, only one (α) of the carboxyl groups of α-ketoglutaric acid was labelled, therefore, the citric acid must have been *asymmetrically* labelled (as indicated in the scheme). This is a reasonable application of the microgeometrical approach to the understanding of organic reactions; however, it leaves, in my opinion, some questions unanswered. Thus, the following two observations do not appear at first sight to be consistent with each other:

(a) if one isolates (labelled) citric acid from a rat liver homogenate, into which labelled carbon dioxide had been introduced, the product is degraded enzymatically—under appropriate conditions—to α-ketoglutaric acid, exclusively labelled in the α-carboxyl group.[32]

(b) if (labelled) citric acid is prepared from levorotatory β-carboxy-γ chloro-β-hydroxybutyric acid and (labelled) sodium cyanide

$$ClCH_2\underset{\underset{COOH}{|}}{\overset{\overset{OH}{|}}{C}}CH_2COOH \xrightarrow{NaC^*N} N\overset{*}{C}.CH_2.\underset{\underset{COOH}{|}}{\overset{\overset{OH}{|}}{C}}CH_2COOH$$

$$\xrightarrow{hydrolysis} HOO\overset{*}{C}.CH_2-\underset{\underset{COOH}{|}}{\overset{\overset{OH}{|}}{C}}-CH_2.COOH,$$

the *same* enzymatic degradation of the citric acid gives α-ketoglutaric acid, which is labelled essentially in the γ-carboxyl group.[33] This appears to indicate that our stereochemical theories are not adequate, and that molecules of the citric acid type retain a memory of their mode of synthesis. If this is true, the approach to chemical and biochemical kinetics, introduced by Polanyi, will once more have proven its usefulness; it seems likely that any solution of this problem will also have to be based on this microgeometrical trend of thought.

NOTES

1. M. POLANYI, *Atomic Reactions*, London: Williams and Norgate, 1932.
2. E. BERGMANN, POLANYI, and SZABO, *Ztschr. physikal. Chem.*, B, 20, 1933, p. 161.
3. POLANYI, and E. BERGMANN, *Naturwiss.*, 21, 1933, p. 338.
4. WOODS, *Brit. J. Exptl. Pathol.*, 21, 1940, p. 74.
5. ROBLIN, LAMPEN, ENGLISH, COLE, and VAUGHAN, *J. Am. Chem. Soc.*, 67, 1945, p. 290.
6. ANDERSON, *Science*, 101, 1945, p. 565.
7. AVAKIAN, MOSS, and MARTIN, *J. Am. Chem. Soc.*, 70, 1948, p. 3075.
8. EMERSON and TISHLER, *Proc. Soc. Exper. Biol. Med.*, 55, 1944, p. 184.
9. McILWAIN, *Brit. J. Exp. Path.*, 21, 1940, p. 136.
10. WOOLLEY, and SHAW, *Science*, 124, 1956, p. 34; *Ann. N.Y. Acad. Sci.*, 66, 1957, p. 649; WOOLLEY, *Proc. Conf. on Neuro-endocrinology*, New York, 1957, p. 127.

11. CRAIG, MOORE, and RITCHIE, *Australian J. Chem.*, 12, 1959, p. 447.
11a. RHEINBOLDT and LEVY, *Chem. Abstr.*, 46, 1952, p. 7551.
12. Cf. PETERS, *Advances in Enzymology*, New York: Interscience Publishers, Inc., 1957, Vol. 18, p. 113.
13. See e.g., MUNIER and COHEN, *Biochim. et Biophys. Acta*, 31, 1959, p. 378; COHEN and MONOD, *Bacteriol. Revs.*, 21, 1957, p. 169; Richmond, *Biochem. J.*, 74, 1960; Meunier, *Compt. Rend, Acad, Sci.*, 250, 1960, p. 3524.
14. VAUGHAN and STEINBERG, International Congress of Biochemistry, Vienna, 1958.
15. HEIDELBERGER, CHAUDHURI, DANNEBERG, MOOREN, and GRIESBACH, *Nature*, 179, 1957, p. 663; Cf. Gordon and Staehelin, *Biochim. Biophys. Acta*, 36, 1959, 351.
16. SHARON and LIPMANN, *Arch. Biochem. Biophys.*, 69, 1957, p. 219.
17. PARDEE and PRESTIDGE, *Biochim. et Biophys. Acta*, 27, 1958, p. 339.
17a. Cf. Naona and Gors, *Compt. Rend. Acad, Sci.*, 250, 1960, 3889.
18. WYSS, RUBIN, and STRANDSKOV, *Proc. Soc. Exp. Biol. Med.*, 52, 1943, p. 155.
19. F. BERGMANN, *Cancer Research*, 2, 1942, p. 660.
20. RIEMSCHNEIDER, in *Advances in Pest Control Research*, New York: Interscience Publishers, 1951, Vol. 2, p. 307.
21. JANSZ, BERENDS, and OOSTERBAAN, *Recueil Trav. chim. Pays-Bas*, 78, 1959, p. 876.
22. COOPER and EDGAR, *Biochem. J.*, 20, 1926, p. 1060.
23. FREUNDLICH and SCHIKORR, *Kolloidchem, Beihefte*, 22, 1926, p. 1.
24. FRIESS, WHITCOMB, DURANT, and REBER, *Arch. Biochim. Biophys.*, 85, 1959, p. 426.
25. WILSON and F. BERGMANN, *J. Biol. Chem.*, 185, 1950, p. 479; WILSON, *ibid.*, 197, 1952, p. 215.
26. PAULING, *Am. Chem. Soc.*, 49, 1927, p. 765.
27. WILSON and F. BERGMANN, *J. Biol. Chem.*, 186, 1950, p. 683.
28. DIKSTEIN, *Biochim. Biophys. Acta*, 36, 1959, p. 397.
29. In the case of acetylcholinesterase, this method has been successfully used by F. BERGMANN and SEGAL, *Biochem. J.*, 58, 1954, p. 692.
30. M. BERGMANN, ZERVAS, FRUTON, SCHNEIDER, and SCHLEICH, *J. Biol. Chem.*, 109, 1935, p. 325. See M. BERGMANN and FRUTON, *Advances in Enzymology*, 1, 1941, p. 63.
31. OGSTON, *Nature*, 162, 1948, p. 963.
32. POTTER and HEIDELBERGER, *Nature*, 164, 1949, p. 180.
33. WILCOX, HEIDELBERGER, and POTTER, *J. Amer. Chem. Soc.*, 72, 1950, p. 5019.

PART TWO

Historical Perspectives

The inquiring scientist's intimations of a hidden reality are personal. They are his own beliefs, which—owing to his originality—as yet he alone holds. Yet they are not a subjective state of mind, but convictions held with universal intent, and heavy with arduous projects. It was he who decided what to believe, yet there is no arbitrariness in his decision. For he arrived at his conclusions by the utmost exercise of responsibility.

P.K., p. 311

6

Kepler and the Psychology of Discovery

ARTHUR KOESTLER

I PROPOSE to concentrate on two related subjects. Firstly, the introduction by Kepler of the concept of physical forces into cosmology; and secondly, Kepler's discovery and subsequent rejection of universal gravity.

Canon Copernicus died in 1543, at the age of seventy. The first printed copy of his book on the *Revolutions of the Heavenly Spheres* was handed to him on his death-bed. It was the only scientific work he had published; its first thirty pages contained a lucid outline of the heliocentric universe. The remainder of the book, more than ten times as long, dealing with the motions of the earth, moon, and planets, is of a supreme unreadability, which made it into an all-time worst seller. Its first edition of a thousand copies was never sold out, and it had altogether four reprints in four hundred years. By way of comparison: Christophe Clavius's textbook, *The Treatise on the Sphere*, had nineteen reprints within fifty years—Copernicus's book, one.

I mention this curiosity because it illustrates the fact that the Copernican theory attracted very little attention on the continent of Europe for more than fifty years, that is, for the next two generations. It was put on the Index only in 1616, that is, seventy-three years after Copernicus's death (and then only for four years pending corrections). The first voice on the continent raised publicly in favour of the Copernican theory was Johannes Kepler's. His *Mysterium Cosmographicum*, published in 1597 (that is, fifty-four years after Copernicus's death) started the Copernican controversy; Galileo entered the scene fifteen years later.

Kepler was twenty-five when he wrote the *Mysterium* and at that time he knew very little of astronomy. He had not started as a scientist, but as a student of theology at the University of Tuebingen in Wuerttemberg; a chance opportunity had led him to accept the post of a teacher of mathematics at the provincial school of Gratz in Styria. In a self-analysis which he wrote at twenty-five, he describes the varied interests of his student years, which ranged from the writing of comedies and of Pindaric lays to compositions 'on unusual subjects such as the resting place of the sun, the sources of rivers, the sight of Atlantis through the clouds, the heavens, the spirits, the genii, the nature of fire, and other things of the same kind'. In this varied menu of preoccupations, in which theological controversies about freedom of will *versus* predestination play a leading part, we find the following remark: 'I often defended the opinions of Copernicus in the disputations of the candidates and I composed a careful disputation on the first motion which consists in the rotation of the earth; then I was adding to this the motion of the earth around the sun *for physical, or if you prefer, metaphysical reasons*'.

I have emphasized the last words because they contain the *leitmotif* of Kepler's quest and can be found *verbatim* repeated in various passages in his works.

Kepler's teacher of astronomy at the University of Tuebingen was a certain Michael Maestlin, one of the few people who had read Copernicus's book and possessed a copy of it, though he himself taught in his textbook the traditional, Ptolemaic, earth-centred universe. This is how Kepler first became acquainted with the Copernican idea. His 'Preface to the Reader' in the *Mysterium Cosmographicum* opens:

'Before I start on my subject proper, I would like to tell you something about what led me to writing this little book and the method by which I proceeded, which, I think, will help you to understand and to get acquainted with my person.

'Six years ago in Tuebingen, when I was eagerly devoted to discussion with the famed magister, Michael Maestlin, I felt already how clumsy in many respects our hitherto prevailing idea of the structure of the cosmos is. Therefore I was so delighted with Copernicus, whom my teacher often mentioned, that I not only frequently defended his views in debates with other students, but also wrote a careful disputation. I thus proceeded, for physical, or if you like this better, for metaphysical reasons, to ascribe to the earth the apparent motions of the sun—as Copernicus does for mathematical reasons ...'

What were these metaphysical reasons?

'My ceaseless search concerned primarily three problems, namely, the number, size, and motion of the planets—why they are just as they are and not otherwise arranged. I was encouraged in my daring inquiry by that beautiful (analogy) between the stationary objects, namely, the sun, the fixed stars, and the space between them, with God the Father, the Son, and the Holy Ghost. I shall pursue this analogy in my future cosmographical work.'

Twenty-five years later, when Kepler was over fifty, he published a second edition of that work of his youth which contains the following note referring to the analogy between sun, fixed stars, and the space between them, with the Father, the Son, and the Holy Ghost:

'I have further dealt with this analogy in Book One of my Epitome and in Book Four of the same work. It is by no means permissible to treat this analogy as an empty comparison; it must be considered by its Platonic form and archetypal quality as one of the primary causes.'

He believed in this to the end of his life; it was one of the three fallacious axioms on which his edifice was built. The other two were his belief in the Pythagorean solids as the invisible scaffolding around which the solar system is built, and the harmony of the spheres governing the planetary motions. But I shall concentrate on the first.

In a letter to Maestlin of the same period, he is a little more explicit:

'The sun in the middle of the moving stars, himself at rest and yet the source of motion, carries the image of God the Father and Creator. He distributes his motive force through a medium which contains the moving bodies, even as the Father creates through the Holy Ghost.'

We notice that here the moving bodies, that is, the planets, are brought in. The Holy Ghost does not merely fill the space between the motionless sun and the motionless sphere of the fixed stars. It is an active force, a *vis motrix*, which drives the planets.

It is my submission that the passages I have quoted reflect the first hesitant emergence of dynamic concepts in cosmology—which, up to Kepler, had been a purely descriptive geometry of the skies, divorced from physics; a point to which I shall return. For the moment I would like to follow the gradual subjection in Kepler's mind of the analogue of the Holy Ghost to the inverse square law.

In the twentieth chapter of the *Mysterium*, he attacks the problem of the mathematical relation between a planet's mean distance from the sun and the length of its year. It was a question which nobody before Kepler had raised—for the simple reason that nobody suspected the existence of physical forces acting between sun and planets. Now, the greater their distance from the sun, the slower the planets move, both regarding angular and tangential velocity. This phenomenon, says Kepler, admits only of the two following explanations:

'Either the souls (*animae*) which move the planets are the less active the farther the planet is removed from the sun, or there exists only *one moving soul in the centre of all the orbits*, that is, the sun, which drives the planet the more vigorously the closer the planet is, but whose force is quasi-exhausted when acting on the outer planets because of the long distance and the weakening of the force which it entails.'

Twenty-five years later, in the Notes to the second edition of the book, Kepler made two revealing remarks. Commenting on the first hypothesis that each planet has its own *anima*, he says: 'That such souls do not exist I have proved in my *Astronomia Nova*.' Commenting on the second hypothesis (a force emanating from the sun), he says:

'If we substitute for the word "soul" the word "force" then we get just the principle which underlies my physics of the skies in the *Astronomia Nova*. . . . Once I firmly believed that the motive force of a planet was a soul . . . Yet as I reflected that this cause of motion diminishes in proportion to distance, just as the light of the sun diminishes in proportion to distance from the sun, I came to the conclusion that this force must be something substantial—

"substantial" not in the literal sense but ... in the same manner as we say that light is something substantial, meaning by this an unsubstantial entity emanating from a substantial body.'

For the rest of his life, Kepler struggled with this new concept of his, the concept of a *vis motrix*, a physical force, without ever coming to terms with it. The main obstacle was *not* Aristotle—*not* that profound appeal which the animism of Aristotelian physics had for two thousand years exerted on the human mind—for in the end he *did* break away from Aristotle. No, the main obstacle was the paradoxical and self-contradictory nature of the concept of force itself.

At first, he was not aware of these difficulties. In a letter to Herwart, he wrote at the time when he was working on the *Astronomia Nova*:

'My aim is to show that the heavenly machine is not a kind of divine, live being, but a kind of clockwork (and he who believes that a clock has a soul, attributes the maker's glory to the work), insofar as nearly all the manifold motions are caused by a most simple, magnetic, and material force, just as all motions of the clock are caused by a simple weight. And I also show how these physical causes are to be given numerical and geometrical expression.'

Having thus defined the essence of the scientific revolution, he tried to visualize that 'moving force' which, emanating from the sun, drives the planets:

'Though the light of the sun cannot itself be the moving force ... it may perhaps represent a kind of vehicle, or tool, which the moving force uses. But the following considerations seem to contradict this. Firstly, the light is arrested in regions that lie in shade. If then, the moving force were to use light as a vehicle, then darkness would bring the planets to a standstill. ...

'This kind of force, just as the kind of force which is light ... cannot be regarded as something which expands into the space between its source and the movable body, but as something which the movable body receives out of the space which it occupies. ... It is propagated through the universe ... but it is nowhere received except where there is a movable body, such as a planet. The answer to this is: although the moving force has no substance, it is aimed at substance, i.e. at the planet-body to be moved. ...

'Who, I ask, will pretend that light has substance? Yet nevertheless it acts and is acted upon in space, it is refracted and reflected, and it has quantity, so that it may be dense or sparse, and can be regarded as a plane where it is received by something capable of being lit up. For, as I said in my *Optics*, the same thing applies to light as to our moving force: it has no present existence in the space between the source and the object which it lights up, although it has passed through that space in the past; it "is" not, it "was", so to speak.'

One feels that Kepler's gropings bring him closer to our contemporary concept of the field than to the Newtonian concept of force. And that, I believe may be the reason why Kepler, having hit on the concept of universal gravity, subsequently discarded it as Galileo was to discard it.

The most striking pre-Newtonian formulation of gravity is in the preface to the *Astronomia Nova*. Kepler starts by refuting the Aristotelian doctrine according to which all 'earthy' matter is heavy because it is its

Kepler and the Psychology of Discovery

nature to strive towards the centre of the world, whereas all 'fiery' matter strives by its nature towards the periphery of the universe and is therefore 'light'. He explains that there is no such thing as 'lightness',

'but matter, which is less dense either by nature or through heat, is relatively lighter ... and therefore less attracted than heavier matter.... Supposing the earth *were* in the centre of the world, heavy bodies would be attracted to it, not because it is in the centre, but because it is a material body. It follows that, regardless where we place the earth, heavy bodies will always seek it.'

And now for the general concept of gravity:

'Gravity is the mutual bodily tendency between cognate [i.e. material] bodies towards unity or contact (of which kind the magnetic force also is), so that the earth draws a stone much more than the stone draws the earth....

'If the earth and the moon were not kept in their respective orbits by a spiritual or some other equivalent force, the earth would ascend towards the moon one fifty-fourth part of the distance, and the moon would descend the remaining fifty-three parts of the interval, and thus they would unite. But this calculation presupposes that both bodies are of the same density.

'If the earth ceased to attract the waters of the sea, the seas would rise and flow into the moon....

'If the attractive force of the moon reaches down to the earth, it follows that the attractive force of the earth, all the more, extends to the moon and even farther....

'If two stones were placed anywhere in space near to each other, and outside the reach of force of a third cognate body, then they would come together, after the manner of magnetic bodies, at an intermediate point, each approaching the other in proportion to the other's mass.'

We have here also the first correct theory of the tides, which in the same passage he explains as a motion of the waters 'towards the regions where the moon stands in the zenith'. Moreover, in a work written at the same time, but published much later—*Somnium*, a dream of the moon (which, incidentally, is the first work of science fiction)—he furthermore postulated that the tides are due not only to the attraction of the moon, but of the moon and sun combined, in other words, that the earth was subject to the attraction of the sun.

But now we come to the paradox. In the *Preface* to the *New Astronomy*, which I have just quoted, Kepler has grasped the essence of gravity and even that it is proportionate to mass—but in the text of this and all subsequent works, he has completely forgotten it. The force which emanates from the sun in the Keplerian universe is *not* a force of attraction, but a *rotational* force which sweeps the planets round in their orbits like a broom that is hinged on the sun. This circular sweeping force is counteracted by the planets' laziness or inertia. Since the sweeping force diminishes with distance, the more distant planets move slower than the nearer ones. Thus the role of gravity and of inertia are reversed in the Keplerian universe. There are no centrifugal and centripetal forces; both forces are tangential. (For the sake of completeness, I must mention, though this is not directly relevant to the issue, that he needs still a third force to account for the ellipticity of the orbits. This is magnetism, which in the aphelion attracts, in the perihelion repels, and thus provides the necessary orbital correction.)

Two questions arise, firstly, what made Kepler drop gravity and, secondly, how was it possible that the Keplerian 'physics of the skies' (that is the sub-title of the *Astronomia Nova*), though it went completely wrong, nevertheless resulted in the three correct laws of planetary motion.

To the first question no direct answer is given anywhere in Kepler's profuse writings. Everything points to an unconscious, psychological blockage, and we may gather hints as to the nature of this blockage in the writings of Galileo, Descartes and Newton. Galileo rejected with disgust Kepler's idea that the tides were due to the moon's attraction, considering it an occult superstition. In the *Dialogue on the Two Principal Systems of the World*, where Galileo's own erroneous explanation of the tides plays a decisive part, he says that Kepler

'despite his open and penetrating mind, lent his ear and his assent to the moon's dominion over the waters, to occult properties [that is, gravity] and such-like fancies.'

Descartes was equally repelled by the idea of a non-mechanical attractive force acting at a distance, and endorsed Kepler's idea that the planets are driven around the sun by a force like a whirlpool. Lastly we have Newton's famous third letter to Bentley:

'It is inconceivable, that inanimate brute matter should, without the mediation of something else, which is material, operate upon, and affect other matter without mutual contact.... And this is one reason why I desired you would not ascribe innate gravity to me. That gravity should be innate, inherent, and essential to matter, so that one body may act upon another, at a distance through a vacuum, without the mediation of anything else, by and through which their action and force may be conveyed from one to another, is to me so great an absurdity, that I believe no man who has in philosophical matters a competent faculty of thinking, can ever fall into it....'

Kepler, Galileo, and Descartes did not fall into that philosophical abyss—their thinking was much too modern, that is, mechanistic, for that. The notion of a 'force' which acts without an intermediary agent and pulls at immense stellar objects with ubiquitous ghost fingers, appeared to them mystical and unscientific, a lapse into that Aristotelian animism from which they had just broken loose. Universal gravity, *gravitatio mundi*, smacked of being a revival of the *anima mundi*. I think the lesson from all this is fairly obvious.

Now to the second question. How did Kepler's incorrect physical assumptions lead him to the correct laws? The answer necessitates a detour.

We remember that Kepler was originally attracted to the Copernican system for 'physical or metaphysical reasons'. His metaphysics remained to the end of his life Pythagorean mysticism. This had two main aspects:

firstly, the conviction that mathematics was the language of God; secondly, the belief that the sun was the hearth of the universe, the source of light, heat, and energy, the watchtower of Zeus transformed into the symbol of God the Father. (Kepler was, of course, acquainted, as Copernicus had been, with the heliocentric cosmology of the Pythagoreans). Aristotle had turned the Pythagorean cosmology upside down: he had expelled God and the First Mover from the centre to the periphery of the universe. Copernicus had gone back to Pythagoras and transferred the centre of the universe back into the sun, but only half-heartedly and incompletely. The Copernican sun was no *primum mobile*; he had neither divine attributes nor did any physical forces emanate from him to drive the planets; nor was he even geometrically the centre of the orbits—the point of reference remained the centre of the earth's orbit; nor did even the planes of the orbits meet in the sun. Kepler found all this highly unsatisfactory, both for physical and for metaphysical reasons. His first moves in his eight-year labours with the *Astronomia Nova* were to make the planes of all orbits meet in the sun and to make the sun the geometrical centre of reference in his system. It is fascinating to work through the tortuous labyrinth of Keplerian geometry, and to see how his *a priori* conviction makes him, quite unconsciously, shift his attention from the radii of epicycles and eccentrics to the radius vector connecting a planet to the sun.

Parallel to this we have the gradual transformation in his mind of the analogue of the Holy Ghost into a physical force. The leitmotive is again his *a priori* conviction that there must be a force coming from the sun which drives the planets, though he is unable to visualize this force and engages in the wildest conjectures. But the details of these do not matter. What matters is his conviction that physical causality permeates the universe. This conviction leads to a series of consequences. First and most important, the *New Astronomy* must be a 'Physics of the Skies', as its sub-title provocatively announces, and this again means that it must provide a workable model of the universe. In the two thousand years from Aristotle to Kepler, this aim was excluded from the astronomer's task. The task was to produce geometrical constructs which resolved the planetary motions into a series of epicycles and eccentrics, a purely fictitious clockwork to save the phenomena and make them conform to the dogma of uniform motion in perfect circles. Ptolemy had used forty circles, Copernicus had been forced to use forty-eight. The accepted method was, when improved observations necessitated readjustments in the orbits, to introduce the required number of auxiliary epicycles and thereby save the appearances.

The great liberating act of Kepler, and the beginning of modern cosmology, was his renunciation of uniform motion in perfect circles. This enabled him to relegate the whole nightmarish machinery of epicycles, of fictitious wheels-turning-on-wheels, into the lumber-room, as he said. But the courage to do this was derived from his revolutionary assumption of physical forces acting between the heavenly bodies, by the fusion in his mind between two hitherto strictly separated disciplines, sky-geometry and earthly physics. In the Aristotelian cosmos, physical forces operated only among the four elements in the sublunary sphere; the motion of the heavenly bodies was not subject to them; they were governed by the criteria of geometrical perfection. Kepler broke down this dualism; by importing, via the Holy Ghost, causality into the skies, it became psychologically possible for him to rid himself of the millenial obsession of uniform motion in perfect circles and the obsessional epicyclitis which it entailed. But the inspiration came from the analogy between the Holy Trinity and the cosmic trinity.

However, we know that the Keplerian dynamics of the skies, though revolutionary, was erroneous; how, then, could it lead to the correct result? The answer is a long and complicated story of intuitive flashes—some right, some wrong—rectified by scrupulous respect for empirical facts, a story which I have tried to retrace elsewhere; but the gist of it is this. Firstly, all previous cosmologies had operated with a single driving force for each planet—angels, spirits, crystal wheels, what you will. Kepler introduced two antagonistic forces—the sun's driving force and the planets' inertia—and this dynamic concept, however rudimentary, directed his search to the relevant problems, enabled him to ask the right questions, and provided him with the clues for his Second and Third Laws.

His original questions had been sterile: why are there six planets instead of twenty or a hundred, and why are they distributed in space as they are? But once he became obsessed with 'sky physics', he hit on the productive questions: (a) what is the relation between a given planet's changes in velocity and its distance from the sun? (b) what is the relation between the various planets' periods and their mean distances from the sun? Out of these two questions came the Second and Third Laws.

The Second Law (which chronologically was the first) was a result of his observation that a planet's velocity at the aphelion and perihelion stands in inverse ratio to its distance from the sun. From this discovery he arrived, by a series of incorrect generalizations and faulty deductions, at the correct result that the radius vector sweeps over equal areas in equal times. The Third Law (which came last), he found after years of trial and error; but the energy for this incredibly dogged effort was again derived from the conviction that a simple law linking period and distance must exist for physical reasons. The First Law, which he found after the Second and before the Third, could only be discovered after his liberation by a shift of his attention to dynamics from the geometrically hypnotic effect of the perfect circle. His solar sweeping force, planetary inertia and magnetism did not, of course, allow him to deduce that the orbit was an ellipse; but it would have been impossible for him to discover that this was the case had he not for physical, or if you like, metaphysical reasons, chosen to operate with a geometry of eccentric polar co-ordinates focused

Kepler and the Psychology of Discovery

on the sun and using the optic equation and the radius vector as the main variables.

To sum up: I have said nothing about Kepler's obsession with the harmony of the spheres and the five Pythagorean solids and have concentrated on a single aspect in his development which I believe to be the one most relevant to the psychology of discovery. It is the fusion between the previously unconnected fields of descriptive astronomy or sky geometry, on the one hand, and physical concepts on the other. Out of this fusion, Kepler's three Laws, the foundation of the Newtonian universe, were born. The fusion has a creative and a destructive aspect. When the two previously unconnected fields or universes of thought are brought into contact, each effects a transformation in the other, breaks down the other's rigid pattern of axioms, dogmas, traditional habits of thought, and at the same time each has a fertilizing influence on the other by showing familiar configurations under a new angle, in a new light. It is a process of *reculer pour mieux sauter*. The observational data on descriptive astronomy on Kepler's desk remained the same, but the relations between the data were transformed, as the bits in a kaleidoscope when shaken up preserve their identity but form a new pattern. The data had previously been related by a pattern of interlocking circles; the shock imparted by the fusion made these vanish and be replaced by an incomparably simpler pattern of vectors symbolizing dynamic forces and tracing six orbits of quasi-concentric ellipses.

Simultaneously, the other field participating in the fusion, physics, was also transfigured. The unmoved movers, spirits and planetary souls as physical causes of the heavenly motions were subjected to the rules of mathematical geometry, lost their independence and freedom and became 'blind', inanimate forces, governed by predictable laws. I have elsewhere tried to defend the view that, in a less dramatic form, the fusion of two previously unconnected frames of reference is an important characteristic of all truly original discoveries and creative processes.

A last remark about Kepler's rejection of universal gravity. He had to reject it as Galileo and Descartes did, because it did not fit his purpose: the construction of a mechanistic, clockwork universe. There exists a type of discovery with delayed effect. Aristarchus's heliocentric system, the theory of impetus of the fourteenth-century Paris School, the finite but unbounded universe of Cusanus, Copernicus's own system, Kepler's formulation of gravity or, in more recent days, Mendel's Laws, all had to undergo a period of incubation before being accepted. One of the illuminating things about Kepler is that the psychological resistance which caused these delays did not come from stupid contemporaries, but can be localized, as it were, in his own mind.

7

The Scientists and the English Civil War

C. V. WEDGWOOD

Some time in the personal rule of King Charles I, as John Aubrey records, the mathematician John Pell was invited to dinner by John Williams, Bishop of Lincoln, 'for the freer discourse of all sorts of literature and experiments.'[1] If Aubrey's chronology is correct the occasion would have been after the Bishop's disgrace, when he was no longer keeping open house in his fine Huntingdonshire mansion of Buckden, where 'the choicest and most able of both Universities came thick unto him'.[2] He may even have been in the Tower at the time, where he continued to entertain as generously as he could.

Pleased with the abilities of John Pell, the Bishop would have offered him a benefice, but the mathematician, with remarkable honesty, said that 'being no Divine and having made the Mathematics his main study' he did not think himself suitable for such preferment. Williams, impressed by this answer but distressed at the rejection of the only form of permanent help that it was in his power to give, broke into a lament on the general lack of patronage for the sciences; 'What a sad case it is that in this great and opulent kingdom there is no public encouragement for the excelling in any Profession but that of Law and Divinity. Were I in place, as once I was, I would never give over praying and pressing His Majesty till a noble Stock and Fund might be raised for so fundamental, universally useful, and eminent a science as Mathematics.'[3]

Though Bishop Williams was never to be in a position where he could influence the Crown to this excellent purpose, the first decades of the seventeenth century did see the growing acceptance by men of perception and learning of the importance of educating the young both in mathematics and in the natural sciences. The King had granted a charter to Sir Francis Kynaston in 1635 for a modern school to be called Musaeum Minervae, where the scholars were to be instructed in physiology, anatomy, astronomy, optics, cosmography, arithmetic, algebra, and geometry, as well as music, fencing, and dancing. Though the older schools still neglected such subjects, the private study of them was increasing, certainly not before it was needed. The practical equipment with which some of the principal ministers of the Crown faced their increasingly complex duties was often lamentable; the Earl of Strafford, governor of Ireland and one of the King's ablest financial advisers, has left occasional scrawled figures among his papers which show him to have been ignorant of the technique of multiplication. A kinsman of his, William Gascoigne, who had become keenly interested in astronomy, in a correspondence with the famous mathematician Oughtred declared that 'he never had so much aid as to be taught addition at school,' and that he 'left both Oxford and London before I knew what any proposition in geometry meant'.[5] Gascoigne's school was no doubt to blame, but he must himself have neglected the opportunities offered by Oxford and London, since Gresham College in London had been offering lectures in geometry since the end of Queen Elizabeth's reign, and Sir Henry Savile had founded the Savilian professorships in geometry and astronomy at Oxford in 1619. It is true that Sir Henry Savile's judgement may not always have been of the wisest, if the story later told by Seth Ward to John Aubrey has any foundation in fact. The geometry professorship at Oxford was open to mathematicians from any part of Christendom, and was accordingly applied for by Edmund Gunter who had been teaching at Gresham College; after watching him perform with his sector and quadrant 'resolving of triangles and doing many fine things', Sir Henry burst out "Do you call this reading of Geometry? This is showing of tricks, man," and so dismissed him with scorn and sent for Henry Briggs from Cambridge.'[6]

But comfortable places in colleges for mathematicians were very limited, and Bishop Williams was only too well justified in his lamentation, in the 1630s, that the branches of knowledge which led to reasonable security in the world were still almost exclusively Law and Divinity. Throughout the seventeenth century it remained usual for those interested in the sciences to pursue a subsidiary profession. The Church was the most usual, though for those principally interested in the natural sciences the medical profession offered a reasonable way of making a living. This accounts for the very high proportion of divines and physicians among the first fellows of the Royal Society.

Among mathematicians John Wallis was remarkable even in his own time for the variety of his gifts. Hearne, enumerating his talents, wrote: 'He was withal a good Divine, and was no mean critic in the Greek and Latin tongues. . . . He had good skill in the Civil Law . . . and 'tis frequently said that he would plead as well as most men, which can hardly be doubted if it be consider'd that he had an extraordinary knack of Sophisticated Evasion.'[7] But if a facility in so many branches was rather exceptional, a reasonable capacity in the exercise of a subsidiary profession was almost an essential for a seventeenth-century man of science. No doubt a talent for Sophisticated Evasion might also come in useful in the quest for patronage. Pell's honesty to Bishop Williams certainly lost him what might have been a safe and suitable livelihood, and twenty-five years later he did indeed take Holy Orders to maintain himself. On the perimeter of the respectable world of science would be found those who, having for one reason or another failed to qualify as divines or physicians got along—and indeed sometimes did very well—out of astrology, alchemy, or journalism.

Though at this stage in the history of scientific thought the era of specialization had not dawned, there was also strong financial pressure to add an accidental element of confusion by compelling thinkers to diffuse their energies. This would not be remedied until there were endowments available to enable men of science to pursue their researches unhampered by the necessity of earning a living.

Though there were vague motions towards a better organization of those concerned with the natural sciences and the new philosophy, nothing much happened in England beyond the friendly association of certain groups around country houses owned by generous patrons. England was a small country where communication and co-operation were not really difficult. As Archbishop Mathew has pointed out, in his study of England under Charles I, 'all serious students were comprised within a circle of correspondents and were to that extent well known.' Thus even William Oughtred, whose humble way of life as vicar of a small Surrey parish caused surprise to distinguished foreign visitors, was under the patronage of the Earl of Arundel, and must have known through him many of the Court virtuosi, certainly Sir Kenelm Digby and, at a later date, John Evelyn.[8] Gascoigne, who pursued his mathematical and astronomical studies in a distant Yorkshire manor house, was probably in touch with Sir Charles Cavendish, the patron of Hobbes, and was known by name to a surprisingly large circle despite his early death.

The King himself was more deeply interested in the arts than the sciences, though he shared the enlightened conviction of most of his fellow rulers in western Europe, that inquiry, so long as it did not interfere with religion or the state, was greatly to be encouraged. He placed the deer in his royal parks at the disposal of William Harvey for his researches into the mysteries of generation. He gave his royal patronage to Theodore de Mayerne who, if as a physician he lacked any great originality of mind, had the perception to rescue from destruction and publish Thomas Moffet's *Theater of Insects*, one of the earliest significant works on entomology. And he encouraged such dilettanti in the sciences as Kenelm Digby and that inveterate inventor of hydraulic machines the Marquis of Worcester.[9]

The deceptive promise of King Charles's Court as a centre of intellectual life was brought to an end by the outbreak of Civil War in the summer of 1642. It would be a delicate question to decide how far the war hampered and how far it stimulated scientific inquiry. In those spheres where war is usually stimulating—namely in the application of science to means of destruction—it was singularly barren. Though Prince Rupert, the King's nephew and cavalry commander, fancied himself both as chemist and mathematician, his hands were too full with day-to-day organization for him to give any time to the experiments which amused his later years. Thirty years on, he and others would be constantly drawing the attention of the Royal Society to the great benefits they might confer by improving the efficiency of fighting ships and cannon,[10] but while he commanded the King's troops his activities in this respect did not go beyond advice and a good deal of personal activity in mining operations—enough at any rate for a courtly poet to call him

At once the Mars and Vulcan of the war.[11]

During the war little or nothing was heard of the numerous inventions of the Marquis of Worcester, not even of the quick-firing devices for muskets and cannon of which his *Century of Inventions* was full.[12] Nor did anyone on either side, in England or in Scotland, resuscitate the burning mirrors and the primitive form of tank which the mathematician Napier of Merchistoun had evolved fifty years before 'for withstanding strangers and enemies of God's Truth and Religion'.[13]

Mathematicians were, however, in demand in other spheres, and in December 1642 John Wallis achieved fame by breaking a Royalist code in an intercepted letter. He remained for the rest of the war the head of Parliament's deciphering department, and used to boast long after the Restoration of King Charles II that he had decoded the King's correspondence taken at Naseby. In later years he would add that he had been careful to omit from his version things which he believed would seriously damage the King's cause. But this was another example of his talent for Sophisticated Evasion; nothing whatever in the King's captured letters escaped the eyes of his enemies.[14] At the end of the war his services to Parliament procured him the Savilian Professorship of Geometry at Oxford, in place of Peter Turner, a worse mathematician, but a better Royalist, who though well into his fifties had volunteered in the King's forces at the very outset of the war.

This volunteering led to some significant losses. William Gascoigne, abandoning his experimental studies in his Yorkshire manor went with other loyal gentlemen to join the Royalist forces before Marston Moor and was among the four thousand dead who were left on that disastrous field. His experiments in the art of flying had achieved no more than two broken legs for the unfortunate boy whom he persuaded to try his method, but he had made interesting progress in perfecting the telescope and his fragmentary correspondence with Oughtred reveals a remarkable mind.[15]

There were other losses of a different kind, of which the most famous (or infamous) was the destruction of Harvey's papers. He had left London hurriedly when the Court fled, and his house was searched after he had gone by ignorant soldiers who took away the manuscript notes of his book

De Insectis. Of this Harvey himself said that 'of all the losses he sustained no grief was so crucifying to him'.[16]

Once he had settled down in the King's headquarters at Oxford Harvey had leisure to pursue the studies that really interested him. It was here, during the war, that he inspected the heart of Lord Montgomery, a nineteen-year-old Irish peer who was active in the King's forces. Owing to an accident in childhood the young man's heart was exposed; he wore a silver plate to protect it. Harvey displayed the obliging young nobleman to the King 'that he might see and handle this strange and singular accident with his own senses, namely the heart and its ventricles in their own pulsation, in a young and sprightly gentleman, without offence to him'.[17] The King concluded the experiment with the melancholy observation to Montgomery: 'Sir, I wish I could perceive the thoughts of some of my nobilities' hearts as I have seen your heart.'[18]

Harvey found numerous friends and disciples in Oxford, and George Bathurst of Trinity College put his rooms at his disposal or rather at the disposal of the sitting hen whose eggs they opened daily in order to observe the progress of the embryo. From the arts of war he held himself aloof. His young friend, Charles Scarburgh, who had been excluded from Cambridge as a Royalist, and whose interests were somewhat divided between medicine and mathematics, became far too enthusiastic an artillery officer for Harvey's liking, and he urged him to 'leave off his gunning' and come back to his proper profession.[19] There was indeed as much need of physicians as of gunners in the war. Harvey was, himself, dragged out of his Oxford preoccupations to attend the King's nephew, Prince Maurice, stricken with a fever during the fighting in the West Country.[20] It was during the Civil War, too, that Richard Wiseman, looking after the wounded at the siege of Taunton, made certain observations on the insensibility of brain tissue which he was to publish many years later.[21]

It needed concentration of purpose and that remarkable serenity of spirit which Harvey enjoyed to resist the interruptions to which all study of whatever kind was subject in Oxford during the war years. In the University city all able-bodied civilians from sixteen to sixty, regardless of their social standing, were required to do one day's work a week on the fortifications (or pay someone else to do it, which was no doubt the usual solution found by men of learning).[22] The trenching and ditching threatened to undermine the Physic Garden founded only twenty years before.

The two Savilian Professors were each infected in different ways with war fever. Peter Turner, Professor of Geometry, had leapt to arms and been almost immediately captured. John Greaves, Professor of Anatomy, took over the running of Merton College, the Warden being in London with the Parliamentary party, to the great annoyance of the sub-warden. He did what he pleased 'of his own strength', and at the end of the war was in trouble with the victorious Parliamentarians for conveying away the money and possessions of the College goods for the advantage of the

Court. He had also lent out goods to courtiers, presumably household stuff which was very badly needed in the overcrowded town. Worst of all, when the Queen lodged for the best part of a year in Merton College, Greaves had been far too familiar with her Roman Catholic suite. He had not only given the Queen's Confessor, Father Philip, leave to use the Library, but he had feasted her chaplains, 'sent divers presents to them, and among the rest an Holy thorn, an excellent instrument of idolatry and superstition at least: and the said Mr. Greaves was observed to be more familiar with these Confessors than any true Protestants use to be.' Finally he had drawn up a petition against the absent Warden, 'inveigled some unwary young men to subscribe it' and so got him voted out of office and replaced by William Harvey.[23]

The Parliamentary Visitors at the end of the war, in the course of purging the University, purged Greaves out of his professorship. He managed, however, to exert some influence in the appointment of his successor, Seth Ward, whose politics were very moderate, and whose attainments and promise were infinitely more notable than those of Greaves. Though it is demonstrably unfair to align progress in the teaching of science during the seventeenth century entirely with the Parliamentary party, there is no doubt that Oxford, which gained Ward, Wilkins, and Wallis under the Puritan dispensation did very well out of it.

Cambridge, behind the Parliamentary lines in East Anglia, had been purged much earlier in the war. A flight of refugees came across to Oxford, where in spite of serious overcrowding they were, on the whole, generously received. They were, for the most part, scholars in law and divinity, though Charles Scarburgh, ejected from Caius College, was one of them. With the support of the energetic Greaves he was taken in at Merton. Seth Ward, his friend, who had joined him in the peaceful days before the war in propagating Oughtred's *Clavis Mathematica* at the University, was ejected at the same time from Sidney Sussex. If he had been resident at Oxford during the war years after his ejection he would probably have been too suspect to be acceptable for the Oxford professorship he got in 1648, so that it was fortunate that instead he sought tranquillity with his old friend and master, Oughtred, at Albury. Oughtred, who received him with the utmost generosity, was himself far from safe. His patron Arundel had very wisely left the country and was far away in Padua. It may therefore have been through Arundel's suggestion that the Grand Duke of Tuscany, Ferdinand II, at this point offered Oughtred a refuge in his dominions and a salary of £500 a year. But Oughtred was over seventy and in any case could not contemplate either changing or concealing his religion.[24] He stayed therefore in Surrey, though Parliament, in control of the county, was purging so-called 'scandalous ministers' and Oughtred as a known Royalist and one who had given more attention to mathematics than to his parish, was an obvious target. Powerful protection was, however, forthcoming to prevent his ejection from his vicarage: or possibly there were Parliamentarians with the discretion to realize that the ejection of a venerable mathematician of international fame would do their cause no good.

Those who had the means to do so, dilettantes or serious thinkers, had mostly left the country. Hobbes had slipped away to Paris. The young John Evelyn, after a momentary thought of joining the King, very sensibly went abroad. John Pell, who had missed his preferment in the 1630s, was fortunate at this time in being called to the professorship of mathematics at Amsterdam, though after the end of the war he was tempted back by Cromwell with a salary of £200 a year. It was curious that having got him home again, the protectoral government proceeded to use him for diplomatic missions. Whether this diffusion of his talent and interruption of his work affected his powers of concentration, or whether he was cursed with a negative and delaying temperament (his own worst enemy) he came to be, in the latter part of the century, a tragic figure of non-fulfilment. 'To incite him to publish anything,' wrote John Collins, 'seems to be as vain an endeavour as to think of grasping the Italian Alps in order to their removal.'[25]

For peace, tranquillity, and the absence of political strain Europe was not far enough away. There may have been some talk, when the younger John Winthrop was in England on the eve of the war in 1641, of a group of experimental philosophers removing to America. But the story is not recorded until about a hundred years later, and the scientists whose names are cited, Robert Boyle and John Wilkins, are not very probable candidates, given the date of Winthrop's visit, when Boyle was only fourteen and abroad in any case, and Wilkins was satisfactorily placed as chaplain to the Elector Palatine.[26] Given Winthrop's interests, and his subsequent relationship with the Royal Society, there is at least nothing improbable in the idea, though the details are evidently wrong.

Robert Boyle and his next brother had been travelling to complete their education, and were caught by the outbreak of the war at Marseilles, whence they retreated judiciously to the theocratic republic of Geneva, where they stayed for two years unable to get any money and living precariously on credit. Though this must have been embarrassing they were well out of the war in Ireland, where their four elder brothers were vigorously and bloodily employed. By the time Robert Boyle got back to England in 1644 the King's cause was visibly declining, but the war looked likely to drag on indefinitely; in Ireland, where some of his patrimony lay, it did in fact continue for another six years.

In London in 1645, against a background of insecurity, political upheaval and the steady penalization of the defeated, began the meetings from which the Royal Society was ultimately to grow. It cannot be truthfully said that these meetings were in any direct way an outcome of the Civil War which provided their unhappy background. All through the 'thirties, in a less deliberate and self-conscious fashion such groups had existed, and the idea of a society of this kind was well known and had distinguished examples in Europe. Yet the conditions of the time gave a negative inspiration; the meetings of the Philosophical Society as its members called it were a silent protest against the irrationality of the times. 'Good God that reasonable creatures, that call themselves Christians, too, should delight in such an unnatural things as war!' wrote the nineteen-year-old Boyle.[27]

He was asked to join their meetings in spite of his extreme youth about a year after they had first begun. The famous account of the first meetings, given some years later by John Wallis, is too well-known to quote but it fixes the earliest date as 1645, the year when the King's defeat became inevitable, and when London had been virtually under martial law and suffering from grave difficulties over food and fuel for nearly three years. Milton, in *Areopagitica*, has left an unforgettable account of London during the heroic months of the war:

> Behold now this vast City: a city of refuge, the mansion house of liberty, encompassed and surrounded with His protection; the shop of war hath not there more anvils and hammers waking, to fashion out plates and instruments of armed Justice in defence of beleaguered Truth, than there be pens and hands there, sitting by their studious lamps, musing, searching, revolving new motions. . . .

It is a stirring vision, but anyone who has lived for long under war conditions knows that, though moments of elation occur, they are flanked by long stretches of drabness and doubt. By 1645 it was becoming apparent that though the war might end, the disturbances and disorders would not, and that a return to stability—especially to moral stability—was far off, and already seemed to many of the older generation an impossibility.

It was then against a background of weariness and gloom that Theodore Haak, a German exile who had been living in England working as a translator and general go-between in intellectual circles for the last twenty years, suggested the formation of a club for weekly meetings. Haak had been long associated with Samuel Hartlib, that indefatigable propagator of useful knowledge, and both of them were in touch with foreign scholars and foreign universities. It was Haak who had been instrumental in getting John Pell his professorship at Amsterdam, and he and Hartlib had, in the days before the civil war, discussed with English virtuosi a plan for an international college. The meetings he now suggested must have seemed at the time a very much lesser substitute.

There was faint, but very faint, hint of grander patronage in the background. John Wilkins was chaplain to the Elector Palatine, then resident in London, and well known to Haak, Hartlib, and the intelligentsia of the capital. Eldest nephew of the King, and dispossessed from his own lands since the beginning of the Thirty Years War, he had come to London, as he claimed, with the intention of reconciling the King and his subjects, but in fact with the tacit hope that the Presbyterian party in the country might regard him as a more suitable King than his uncle.[28] The Elector's patronage was not very lavish because he had not the means, but he was genuinely interested and had given a good deal of help and encouragement to Wilkins, who dedicated to him in 1648 his *Mathematical Magic, or the Wonders that may be performed by Mechanical Geometry.*

At the meetings of the club 'we barred all discourse of divinity, of state affairs, and of news, other than what concerned our business of philosophy', Wallis recorded some years later.[29] The prohibition was essential. Although the political opinions of the earliest members were most of them moderate, inclining towards the current brand of English Calvinism (few of them after the Restoration were to find any great difficulty in rejoining the Episcopal Church in which Wilkins and Ward later became Bishops) the situation in London was, during these first years, so tense and so troubled that only a strict adherence to the 'no politics' rule would eliminate all possible friction.

Thomas Sprat, twenty years later, summed up the situation in his History of the Royal Society: 'Their first purpose was no more than only the satisfaction of breathing a freer air, and of conversing in quiet with one another, without being engaged in the passions and madness of that dismal age.... For such a candid and impassionate company as that was, and for such a gloomy season, what would have been a fitter subject than Natural Philosophy?'[30]

The young Boyle, returning to England after the regulated calm of Calvinist Geneva, was shocked and distressed by the violence and changeability that he found—'they esteem an opinion as a diurnal, after a day or two scarce worth the keeping.'[31] He built his hopes on his new friends as on a rock. 'The corner-stones of the *Invisible* or (as they term themselves) the *Philosophical* College do now and then honour me with their company,' he wrote, 'men of so capacious and searching spirits that school philosophy is but the lowest region of their knowledge ... persons that endeavour to put narrow-mindedness out of countenance.'[32]

The story of the evolution of the Royal Society has been told so often that it has become hackneyed. It was, of course, in one sense a perfectly natural development and a mere imitation of the sort of learned bodies that had already come into being abroad and were to continue to do so throughout the century. But the Accademia dei Lincei, Accademia del Cimento, the Académie des Sciences and others owed their being to powerful official or individual patronage and existed in countries which, if not precisely at peace, were at least held in a more or less static framework of religious and political convention. The Royal Society in its embryo stage, on the other hand, was the outcome of individual co-operation among men of science in a time of almost complete political and religious disintegration.

To 'endeavour to put narrow-mindedness out of countenance' in such an epoch, though it was not unique, was none the less heroic. The insistence of the members of the Invisible College on the secular character of the knowledge they sought, implicit in the refusal to discuss divinity, was really significant. For it did not emanate from men who had turned their backs on religion, but rather from religious men who were ready to separate belief and doctrine from inquiry. Thus 'Mr. Hobbes the Atheist' remained an unusual and slightly shocking figure among seventeenth-century scientists, and those who met in the Invisible College in the late 1640s were men of rather more than merely conventional religious faith. The devout Boyle was to give a great deal of his life and thought to the propagation of the Gospel among the heathen and none of them was indifferent to the ultimate shape that religion would take in their much-tried country. It was thus by implication an expression of a profound faith in the power and usefulness of knowledge that led these men to dissociate themselves for the prosecution of their work from the political and religious disputes in which they were all to some extent 'engaged'. Proud of the achievement, for he was always pleased with the triumphs of learning, John Aubrey was later to write 'Experimental Philosophy first budded here and was first cultivated by these virtuosi in that dark time.'[33] But he did not altogether realize all the implications of that very remarkable achievement. It was neither an escape nor a revolt from current religious beliefs or political doctrines; but it was an emphatic statement that the pursuit of useful knowledge had a right to separate existence.

Meanwhile the oppressed, defeated and exiled, mostly of an older generation, found some solace in the return to their old interests. Hobbes's patron Charles Cavendish, back in England and much impoverished, went on with his mathematical speculations, but to what effect will never be known, since his executor's wife sold all his notes for wastepaper.[34] Kenelm Digby, in exile in Paris, returned to his chemical experiments, but struck the critical young John Evelyn as nothing but an 'arrant mountebank' trying to palm off on him a quack cure for indigestion.[35] In England, in prison in the Tower or released on bail, the Marquis of Worcester collected his *Century of Inventions* and looked sourly towards the experimental station he had made at Vauxhall, which had been taken over, without much success, by the Cromwellian government.

Prince Rupert, in the intervals of commanding a fleet of English privateers and earning his living as a soldier of fortune with the French, the Savoyards, and the Austrians, associated himself for a time with the work of his gifted sister Elizabeth, the disciple and correspondent of Descartes, but later, reverting to the technical problems which always interested him more, worked out in association with Wallerand Vaillant the method of engraving to which Evelyn was to give the name mezzotint.[36]

It was in exile at Maastricht that Sir Robert Moray, after spending half his life as a professional soldier, became an enthusiastic amateur of the natural sciences, gave himself up for long months to the study of chemistry, and acquired the knowledge which was to make him a moving spirit in the ultimate foundation of the Royal Society. There was moreover, a younger generation of nobles and courtiers coming on, better equipped to appreciate and understand the new advancement of learning. Charles II, and the Duke of Buckingham less attentively, studied mathematics with Hobbes. As a prisoner of Parliament in England the young Duke of York had had instruction in mathematics from Oughtred's pupil Jonas Moore.

Though the outburst of gaiety which followed the Restoration showed that the repressions of the 1650s had not suited all temperaments, there were certainly some among the younger generation who had profited by them. 'Persons of Quality, having no Court to go to, applied themselves to their studies,' wrote an observant French visitor in 1662, 'some turning their heads to Chymistry, others to Mechanism, or Natural Philosophy; the King himself has been so far from being neglectful of these things that he has attained to so much knowledge as made me astonished, when I had audience of His Majesty . . . the English Nobility are all of them learned and Polite.'[37]

The writer, Monsieur de Sorbière, was no doubt somewhat carried away by his reception, for he was squired about everywhere by Sir Robert Moray and introduced principally to his virtuoso friends, in or out of the nobility. But it was certainly true that the generation who had lived through the war and its aftermath could look forward, if not to the generous provision that Bishop Williams had once spoken of, at least to an epoch of expansion and encouragement.

NOTES

1. *Aubrey's Brief Lives*, ed. Oliver Lawson Dick, London, 1949, p. 230.
2. JOHN HACKET, *Scrinia Reserata*, London, 1692, Part II, p. 32.
3. AUBREY, loc. cit.
4. RYMER, *Foedera*, XIX, pp. 638–41; see also Weld, *History of the Royal Society*, London, 1878, I, pp. 19–23.
5. RIGAUD, *Correspondence of Scientific Men*, Oxford, 1841, I, p. 84.
6. AUBREY, p. 268.
7. HEARNE, *Remarks and Collections*, ed. Dobell, Oxford, 1885, *seq*, I, p. 198.
8. DAVID MATHEW, *The Age of Charles I*, London, 1951, p. 243; Aubrey, p. 223.
9. At this date Worcester was known as Lord Herbert of Raglan, his father being still alive; in the Civil War he became Earl of Glamorgan, and on his father's death Marquis of Worcester. In the interests of simplicity rather than precision I have called him Worcester throughout.
10. A. R. HALL, *Ballistics in the Seventeenth Century*, Cambridge, 1952.
11. ANON., *Elegy on Prince Rupert*, London, 1682.
12. DIRCKS, *Life of the Marquis of Worcester*, London, 1865, pp. 462 ff.
13. Napier never printed these inventions, thinking them too dangerous, but sent the MS. to Anthony Bacon, brother of Francis.
14. HEARNE, VIII, p. 394.
15. AUBREY, p. 210; RIGAUD, I, pp. 83, 84, 87.
16. AUBREY, p. 128.
17. HARVEY, *Works*, ed. Sydenham Society, pp. 382–3.

18. The anecdote occurs in a footnote under 'Mount Alexander, Earl of,' in G.E.C.'s *Complete Peerage*. Montgomery married, raised a family, and was created Earl of Mount Alexander for his services in the Civil War.
19. AUBREY, p. 129; Aubrey says merely that Scarburgh 'marched up and down with the Army', but Harvey's reference to 'gunning' and Scarburgh's mathematical studies have led me to assume that he was in the artillery.
20. WARBURTON, *Prince Rupert and the Cavaliers*, London, 1849, II, p. 307.
21. LONGMORE, *Richard Wiseman*, London, 1891.
22. STEELE, *Tudor and Stuart Proclamations*, I, No. 2433.
23. *Register of the Visitations of the University of Oxford from 1647 to 1658*, ed. Montagu Burrows, *Camden Society*, New Series, 29, London, 1881, pp. 252–3, 283.
24. AUBREY, p. 224.
25. RIGAUD, I, pp. 196–7.
26. SIR HENRY LYONS, *The Royal Society*, Cambridge, 1944, p. 7.
27. BOYLE, *Works*, I, pp. xxvii.
28. For the curious behaviour of Charles Louis Elector Palatine, during the Civil War, see the present writer's *The King's War*, pp. 589–90.
29. WELD, I, p. 36.
30. SPRAT, *History of the Royal Society*.
31. BOYLE, *Works*, I, p. xxiv.
32. ibid., loc. cit.
33. AUBREY, p. 238.
34. AUBREY, p. 58.
35. Evelyn's Diary, 17 November 1650.
36. See a recent article (1960) by Orovida Pissarro in the Publications of the Walpole Society.
37. SORBIÈRE, *A Voyage to England*, 1662, pp. 32–3.

8

Vibrating Strings and Arbitrary Functions*

J. R. RAVETZ

1. *Introduction*

JUST about two centuries ago, Euler, Lagrange, d'Alembert, and Daniel Bernoulli were engaged in a controversy over a problem in mathematical physics, best described by the title above. Since then the 'Vibrating String' controversy has enjoyed a certain notoriety, but it has never become a 'classic' controversy in the history of science. One reason for this may be that its structure was so decidedly unclassical, for in the case of this controversy even hindsight cannot furnish us with immediate information on which side was 'right'. Indeed, it is a task of some difficulty to demarcate the sides: at no time were there less than three antagonists, and the controversy flourished in an atmosphere of mutual incomprehension. The 'crucial experiment' which serves as the climax of a classic scientific controversy played a most peculiar role here: it was performed (as a mathematical theorem) at the height of the controversy, but no one noticed it, and it had to be rediscovered by Fourier forty years later.

Thus the Vibrating String controversy was a mess. But it is no less instructive for that. For the first time, problems of mathematical physics were the subject of debate on methodological rather than metaphysical grounds. What the controversy lacked in decisiveness, it made up for in the depth of the problems discussed, and (once we understand the patterns of thinking) in the clarity of the distinction between mathematics, physics, and that something which lies between.

* Bibliographical note: The basic bibliographical source for the Vibrating String controversy is J. Burkhardt, 'Entwicklungen nach oszillierenden Functionen . . .' *Jahresbericht der Deutschen Mathematiker-Vereinigung*, 10, 1908, pp. 1–1800; the controversy is dealt with in the first fifty pages of this. Discussion of some technical and political aspects of the controversy, especially those relating to Euler, is given by C. Truesdell, in the Editor's Introduction to Euler's *Opera*, Series II, Volume XIII, Lausanne, 1955.

I have drawn on both sources in places too numerous to mention.

My references to eighteenth-century sources will include only information relating to the most accessible editions. All the Euler papers are found in his *Opera*, Series II, Volume X, ed. F. Stüssi and H. Fabre, Berne, 1947; I give only their page numbers in that volume and their index number in the catalogue of Eneström (*J.D. M.-V., Ergänzungsband IV*, 1910–13).

At the base of the disputed mathematical structures was a well-known physical phenomenon: a string under tension, made to vibrate. Three features of the vibration called for theoretical explanation. The first is that the vibrations are *regular* and *isochronous*: the string vibrates as a whole, and emits a musical sound, not a dissonant noise, and moreover this sound persists unchanged until the vibrations die away completely. The second is that the tone emitted is independent of the manner in which the string is put into vibration (thus, either plucking or bowing a violin). Finally, there are the phenomena of overtones: a string will resonate to musical tones whose frequency is some sub-multiple of its natural frequency; and moreover the trained ear will detect these overtones in the sound emitted by a vibrating string.

We shall see that a theory basing its structure directly on these phenomena was rather slow in coming, and even then was not the most successful. But all the participants in the controversy were aware of the great gap between a mathematical theory and a directly 'physical' explanation. If nothing else, they all used (in different ways) the assumption that the motions are 'infinitely small', and left unexplained the damping action of the air. The differences in approach revealed themselves in terms of the relation which such an abstract theory was considered to have to the experimental phenomena: whether it should explain all (Bernoulli), some (Lagrange and Euler) or none (d'Alembert).

Given this sketch of the physical and methodological background to the problem, we can now appreciate the set of questions on which the controversy turned. These were: *what is the mathematical formalism most appropriate to the problem?* and, *what are the proper limits on the interpretation of the formalism, from the points of view of mathematical rigour and physical plausibility?*

The two competing formalisms were the following. Generalizing from the oscillatory character of the phenomenon, one may give the general equation of the string in motion as

$$y(x,t) = a_1\sin(\pi x/l)\cos(\pi t/T) + a_2\sin(2\pi x/l)\cos(2\pi t/T) + \ldots \quad (1)$$

an infinite series, with undetermined coefficients $a_1, a_2, \ldots$. The first term of the series explains, qualitatively at least, what we see when we look at a thick string (say, that of a 'cello) in vibration. For any given time, the string takes the form of a smooth curve with one arch, fixed at the endpoints $x = 0$ and $x = 1$. On the other hand, each point of the string oscillates up and down with a simple harmonic motion; and indeed the motion of the whole string is like that of a sine curve partaking as a whole of simple harmonic motion. The extra terms serve to explain the phenomena of harmonics, and also to allow more general initial configurations of the string.

The other formalism is a deduction from the equations describing the motions of a small (infinitesimal) portion of the string. These can be expressed as

Vibrating Strings and Arbitrary Functions

[Vertical acceleration of a portion] proportional to [curvature of that portion], (2)

or, in symbols,

$$d^2y/dt^2 = c^2 \, d^2y/dx^2. \quad (3)$$

The 'most general' solution of this equation is

$$y(x,t) = \tfrac{1}{2}[F(x + ct) + G(x - ct)], \quad (3a)$$

where F and G are arbitrary functions. If one assumes that the string is fixed at the endpoints ($y(0, t) = y(l, t) = 0$), has initial velocity zero, and initial displacement $f(x)$, then the solution is

$$y(x, t) = \tfrac{1}{2}[f(x + ct) + f(x - ct)]. \quad (4)$$

Changing the argument of the function from x to $x \pm ct$ produces a *shift* in the graph of the function. Thus if $f(x)$ has the following graph on the interval $0 \leqslant x \leqslant l$:

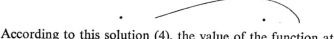

then, for $ct = \tfrac{1}{4}l$, $f(x + ct)$ will look like

and $f(x - ct)$ will look like

According to this solution (4), the value of the function at $ct = \tfrac{1}{4}l$ is the *average* of these two graphs, or

We see immediately that for this solution to be of any use, the function $f(x)$ must be defined outside its original interval. Since this extension of the function was one of the points of controversy, I leave it for later.

A simplified scheme of the arguments is as follows: d'Alembert argued for the 'functional' solution on grounds of mathematical generality, and argued for a more restricted interpretation of the initial $f(x)$ than Euler's on grounds of mathematical rigour. Euler argued for the most general interpretation of $f(x)$ on grounds of mathematical generality, and physical fruitfulness. Daniel Bernoulli argued for the 'trigonometrical' formalism and against the 'functional' one on grounds of physical plausibility. Lagrange came into the argument ten years after it began, espoused a solution of the same generality as Euler (but derived by different means) and agreed with d'Alembert's strictures on Euler's own solution. Needless to say, each participant offered counter-arguments against the claims of opponents, although it was never admitted explicitly that an antagonist was engaged in a different sort of activity from one's own. The controversy (which did not rage continuously) eventually died out as d'Alembert and Bernoulli retired from the field. Euler and Lagrange used their common views as the necessary basis for new advance in mathematical physics. A mathematical justification of their approach was given by Monge, in his 'geometrical' theory of partial differential equations. By the late 1760s the embers of the controversy were barely glowing; but it flared up again forty years later when Fourier achieved what Bernoulli wanted to but could not (and what Lagrange could but did not want to), and set mathematical physics on a new path.

2. The First Phase: the Statement of Positions

(a) TAYLOR

Like any other historical episode, this one has a pre-history. The only event in this pre-history with which we need concern ourselves is Brook Taylor's solution of the Vibrating String problem in his *Methodus Incrementorum* of 1715. This solution remained unchallenged for thirty years; indeed it was one of the few really successful attacks on a problem in continuum mechanics in the first half of the century. Moreover, it remained in the physical literature as *the* solution through to the end of the century. All three of the early entrants into the controversy based themselves on Taylor's work, and so it is worth our while to analyse it.

Although Taylor's mathematical arguments are difficult to follow, the structure of his assumptions and results is quite clear. First, an eminently plausible analysis (the one still common) yields the assumption that at each point of the string

$$\text{Vertical Restoring Force} \propto \text{curvature} \times \text{tension} \qquad (5)$$

The relation of curvature to displacement at each point is derived from the 'regularity' assumption:

The cord assumes a straight-line shape once in each half-vibration (6)

Some infinitesimal geometry based on this yields

$$\text{Displacement} \propto \text{curvature} \qquad (7)$$

The isochronous property of the motion results from combining (5) and (6) to yield

$$\text{Restoring Force} \propto \text{displacement} \qquad (8)$$

This is the defining relation for pendulum motion; the idea of studying oscillating motions *starting* from an analogy with the pendulum was a fruitful one, and was exploited by the Bernoullis, Euler, and others. Taylor was quite explicit about the mathematical assumptions underlying his treatment. Thus, from the assumption

The inclination of the tangent to the curve is infinitely small (9)

he could simplify relation (7) to yield a sine curve as the shape of the string at any instant.

Vibrating Strings and Arbitrary Functions

He recognized the problem of the plucked string (where the initial shape is nothing like a sine curve) and tried to prove that any other initial shape is quickly deformed into a sine curve. Finally, he was able to use his proportionalities and sine-curve shape to derive the still-accepted result that the period of vibration is proportional to $[LN/P]^{\frac{1}{2}}$, where L is the length, N the weight and P the tension of the string.

(b) D'ALEMBERT

Some thirty years later, d'Alembert announced that the vibrating string admits curves more general than Taylor's 'companions of the cycloid', and proceeded to derive a more general solution.[1] Quoting Taylor, he used only the assumption (5), and none of those following from (6) and (7). From this less stringent assumption he derived the functional solutions (3a) and (4). As Bernoulli pointed out a few years later, this is generality bought on the cheap (these were not his exact words). For, if one throws away half the information about the physics of a situation, it is not difficult to get a solution more general mathematically. Then one must face the question of the relevance of the 'general' solution to the actual motions of a vibrating string.

The question must be faced, however, only if one is seriously interested in the actual motions of a vibrating string. And it is abundantly clear from his pair of papers that d'Alembert did not have such an interest. The papers are mainly devoted to the mathematical exploitation of a mathematical achievement of the first order: the derivation of the functional solutions (3a) and (4) from the differential equation (3). Deriving (4) from the end-conditions is by no means trivial, and d'Alembert could find many interesting mathematical excursions on the basis of the derivation. Indeed, in making the derivation he found it necessary to make certain restrictions on the function of $f(x)$. These are that $f(x)$ should be *periodic*: $f(x + 2l) = f(x)$; and that (for zero initial velocity), $f(x)$ should be *odd*: $f(-x) = -f(x)$. From these conditions we see that the function is really defined on the whole line $-\infty < x < +\infty$, and so the difficulties of constructing $\frac{1}{2}[f(x + ct) + f(x - ct)]$ for large ct, mentioned in §1, are solved. Such an 'odd and periodic' function is, of course, $f(x) = \sin(n\pi x/l)$, but d'Alembert was concerned to show that one could construct an infinite class of more general functions satisfying these same conditions. It hardly needed saying, at this stage, that a function so constructed, defined on the interval $-\infty < x < +\infty$ should have a single *law* (not necessarily expressible in an equation) for its definition. I hope that this condition appears natural, for it was eminently natural to d'Alembert, and without it much of his work would have been meaningless.

(c) Euler

Now we have a head-on collision. In a paper published shortly after,[2] Euler re-derived the equation, and the functional solution, formally identical with d'Alembert's. In this he announced, as a natural fact, that the initial-configuration function $f(x)$ can be *any* function defined on $0 \leqslant x \leqslant l$, and that it is then *extended* to be odd and periodic by the piecing together of bits of function on successive intervals of length. Thus if the original $f(x)$ looks like

we extend it to $-l \leqslant x < 0$ by the 'odd' condition $f(-x) = -f(x)$, and have

We can extend it to $l < x \leqslant 2l$ by the 'periodic' condition, and get

and so on, as much as we want. There is nothing to stop one from making such a mathematical construction; the only question is, what does one have at the end? Clearly, not a *function* in the sense of d'Alembert, or even in the sense of Euler's own '*Introductio in Analysin Infinitorum*' of 1748, where one finds the 'classic' eighteenth-century definition of a function as an equation between variables. Indeed, Euler made it plain that even the 'initial-configuration' $f(x)$ need not have a 'formula' or a 'law', but could be a curve drawn by hand.

Considered as a piece of mathematics, Euler's paper is slight compared to d'Alembert's. The possibility of making 'more general' curves is not supported by any construction, the fine points of the behaviour of the string (arising from symmetries in the initial configuration) are given only a brief mention, and none of the mathematical refinements of differential-equations theory are discussed. Once he has the wave equation, Euler derives its solution, gets the 'odd and periodic' conditions for the extension, discusses the simpler features of the motion, and exhibits the 'trigonometrical' solution (1) as a special case.

But Euler was not here doing mathematics; rather was he doing 'rational mechanics'. The background to his paper lies in his earlier attempts to set up a fundamental theory of the motions of elastic bodies,[3] independent of the 'pendulum' assumption (8) mentioned above. It is most probable that he saw the possibility of the differential-equations techniques developed by d'Alembert in an earlier essay, and related these to a new analysis of the dynamics of flexible bodies. All the internal evidence indicates that his paper was written independently of d'Alembert's.

Thus it is plausible to suppose that when Euler derived the functional solution (4), he would examine it with a view to giving it the least restrictive interpretation possible. And here the properties of the 'extended' $f(x)$ provide a powerful result. For one finds that the period of the vibration is independent of the shape of the initial configuration: thus one has at time $t + 2l/c$

$$\begin{aligned} 2y(x, t + 2l/c) &= f[x + c(t + 2l/c)] + f[x - c(t + 2l/c)] \\ &= f(x + ct + 2l) + f(x - ct - 2l) \\ &= f(x + 2l + ct) + f(x - 2l - ct) \\ &= f(x + ct) + f(x - ct) = 2y(x, t); \end{aligned}$$

the curve returns to the same configuration every $2l/c$ units of time. Thus any 'extended' $f(x)$, so long as it were periodic, would give the isochrony result for the vibrating string. The parameter c is clearly independent of the choice of $f(x)$; from the theory one gets a value for it agreeing with Taylor's.

Now one can look at the original $f(x)$, defined on $0 \leqslant x \leqslant l$, and ask why it should be restricted to be part of a function odd and periodic. The formalism does not demand it, and the physics of the situation (initial configuration assigned at pleasure) indicate otherwise. Indeed, viewed as an idealized description of a physical system, the initial function $f(x)$ should be subject to *no* restrictions: and it was this widest possible interpretation which Euler gave to it.

Since so many later arguments centred on this issue of 'generality', it is worth making some clarification without which the older views on functions, involving a gradation through 'algebraic', 'transcendental', 'defined by a continuous law' to 'drawn by hand', seem unnecessarily antique. The point of view of the most modern branches of mathematics is that classes of functions are defined by characterizing certain sets of points, and from the 'smoothest' to the 'roughest' is only a matter of degree. However, the continued existence of distinct *species* of functions (in practice if not in theory) is shown by the difference in approach between 'real variable' and 'complex variable' theory and their generalizations. What makes this paper of Euler's significant for the history of pure mathematics is that the conception of 'greatest generality' reappeared in his work for the first time since Barrow's 'Lectiones Geometricae' were rendered obsolete by the calculus. And it was this conception (although not his opinions concerning it) which led, through Fourier, to the Lebesque integral and modern abstract mathematics.

The first note of controversy was sounded in 1750: d'Alembert submitted a letter to the Berlin Academy simply stating that Euler's solution was false. The letter was not printed, but its effectiveness was not reduced. D'Alembert gave a hint of his criticism in a paper published in the Berlin Memoires of 1750;[4] he reminded readers that $f(x)$ *must* be odd and periodic, and stated that unless the initial configuration is a part of such a function, the problem cannot be solved. For the next ten years, Euler was left to wonder what arguments d'Alembert had up his sleeve. He did not challenge d'Alembert, for the Berlin Academy was then still involved in the unsavoury 'least action' controversy, and one did not go looking for trouble.

(d) BERNOULLI

The opening phase of the controversy was completed by the appearance of Daniel Bernoulli's two papers on vibrating strings.[5] Bernoulli writes as an honest physicist, somewhat bewildered by the mathematical pyrotechnics of Euler and d'Alembert. More than once he says that this is beautiful mathematics, but what has it to do with vibrating strings? In his paper, he is concerned to show that one can get results of the same order of generality without throwing away the physical problem with which one started, without the very abstract mathematics of functions of $x \pm ct$, and with mathematical tools which really show what is going on. These moving graphs of function do not seem to reflect any physical process in the vibrating string; it was not until after the 'wave equation' (3) had been applied to wave phenomena that the vibrating string could be seen as a special case, called 'standing waves'. Moreover, he challenged the assumption that there is an $f(x)$ 'more general' than one given by a trigonometrical expansion. All he said in these papers was that he was not satisfied on this point; it is probable that he thought d'Alembert's constructions of more general curves not worth puzzling out, and did not take seriously Euler's class of curves 'drawn by hand'.

Bernoulli's positive arguments rested on a physical principle, that of superposition of independent vibrations of the string. This has become such a commonplace (Euler and d'Alembert saw it as a *mathematical* feature of the solution of the wave equation) that one may wonder that it needed 'discovering' at all. But as a physical principle it was not at all obvious; fifty years later Lagrange, in his *Mécanique Analytique* (Part II, Sect. VI, Para. 59) denied its validity. The mathematical description of superposed motions, as in equation (1), makes the whole thing look too trivial. Bernoulli started from the physical phenomena of multiple sounds: overtones. By his time it was known that a string tuned to a certain pitch will resonate to a sound an octave lower, and also that any musical sound is composed of a fundamental and overtones with frequencies 2, 3, 4, ...

Vibrating Strings and Arbitrary Functions

times that of the fundamental. How could one reconcile this with the Taylor theory of the string's vibrations, where the curvature (and acceleration) of a part of the string is simply proportional to its displacement from equilibrium?

The physical solution of the difficulty was to imagine the more frequent vibrations as being *relative* to the fundamental. Thus if the natural shape of the string is

then to get the octave, one adds a displacement to this one: and higher-order harmonics are similarly superposed. That is, the moving curve for the fundamental vibration is considered as the *axis* for the moving curve for the octave, and so on. The mathematical difficulties of this procedure are avoided by the assumption that all vibrations are infinitely small; thus the difference between arc length and abscissa can be ignored.

For his critique of the Euler–d'Alembert approach, Bernoulli had two sorts of arguments. The first was to show that all the qualitative features of the string's vibration could be obtained by considering very straightforward classes of curves such as

$$f(x) = \alpha \sin(\pi x/l) + \beta (\sin 2\pi x/l) \qquad (10)$$

and the various symmetry properties come out with next to no calculation. Then he attacked the Euler–d'Alembert approach on physical grounds. Much of his argument can be interpreted (and, it seems, was by his adversaries) as quibbling over words: regular, isochronous. But Bernoulli was arguing in the framework of a physical theory which no one had up to then challenged: the simple vibrations of a string are isochronous and regular. Now, consider an initial configuration of the form (10) with β much less than α, giving a curve of the form of the preceding sketch. Does it have a simple or a compound vibration? Patently, the vibration is compound. But the Euler–d'Alembert interpretation forces one to say that it is simple, and of the same period as that of $\alpha \sin(\pi x/l)$. For if they were to admit that this vibration is compound, they would have to admit that any non-sinusoidal curve yields a compound vibration. Then they would not be able to speak of 'the' period of the oscillation of a non-sinusoidal curve, and moreover they would be forced to admit the necessity for a trigonometric decomposition of a 'more general' curve. Thus one must either accept Bernoulli's arguments in principle, or jettison all the agreed physical facts of the vibrating string.

Bernoulli devoted considerable space in the papers to a study of the oscillations of systems of several particles, showing how the superposition of independent vibrations came naturally from the basic equations. Looking back at his paper, one sees (too clearly) its great weakness: the absence of a demonstration that any function $f(x)$ has a trigonometric expansion, a demonstration which could be given only through an evaluation of the coefficients. I think that there are two related reasons why Bernoulli didn't provide this. First, that he could not, and second that he did not want to. For analytical devices of great sophistication were required to disentangle the formalisms one got from an n-particle analysis, to exhibit the coefficient formulae: even Euler's notations were nowhere near sufficiently developed for this. Moreover, such a result would not interest Bernoulli, unless it could be obtained very cheaply: his concern all along was to describe and explain the accepted physical phenomena with the minimum of mathematical elaboration. It is clear from his writing that although he was thoroughly proficient in mathematics, he did not have the mastery necessary for the invention (or even assimilation) of the new techniques required for a mathematical attack on the problem. Considering that since 1733 he had been Professor of Anatomy and Botany at Basle, his achievements in this line are, to say the least, creditable.

3. *The Second Phase: Criticism and Recrimination*

With the critical passages of Bernoulli's paper the second phase of the controversy was opened. I shall not attempt a complete chronicle of the arguments, for most of them are illuminating only for a deeper study of the controversy. D'Alembert was, of course, in the most comfortable position: he could be as priggish as any pure mathematician in controversy with physicists. Bernoulli could base himself on the physics of the problem, but to attack his adversaries on the issue of 'generality' he had to stray from his territory into theirs. Euler was exposed to cross-fire from both directions, and perhaps because of this his ideas developed more than those of the others.

(a) EULER

Euler's first reply to his critics[6] appeared in the *Berlin Mémoires* directly after Bernoulli's papers. In it there is an interesting pair of assertions. Dealing with the possibility that vibrations, however irregular at the beginning, are deformed into regular ones, he compliments Bernoulli on his ingenious idea that this may occur as an effect of air resistance. But, says Euler, the whole problem has been treated on the assumption that there is no resistance; and it is within this unreal framework of assumptions that the arguments must be carried on. (Later he lists the assumptions, and remarks that they are strong abstractions from reality.) Yet in returning to

Vibrating Strings and Arbitrary Functions

the fact of non-sinusoidal initial conditions, he says that the (real) cord does move, and the movement, being real, must be determinable. One might ask Euler to make up his mind whether he is talking about the real physical world (with Bernoulli) or a mathematical one (with d'Alembert), but such a request would be misguided. Euler was working in the no-man's-land between the two, and the battle for the existence of this territory had been won for him by Galileo.

In his second paper of self-defence,[7] Euler gave more consideration to d'Alembert's thesis that only 'analytic' curves $f(x)$ are acceptable as initial configurations. For this he had a double answer: if these are fallacies they will appear soon enough (a good maxim, in view of the state of analysis in the mid-eighteenth century); and the study of partial differential equations is quite different from that of 'the calculus' of functions of a single variable. Whereas in the latter, all operations are performed on one or another analytical expression, in this new branch of mathematics absolutely arbitrary functions must enter. Neither of these arguments is of overriding force, and we will see how Lagrange's paper (published before this one was published, but almost certainly after it was written) made a great impression on Euler.

There are two other important developments here. First, Euler no longer attempts to 'construct' the general solution from the equation: he exhibits it, and shows that it satisfies the equation. Without this greater freedom of approach, it would have been impossible for Euler to find solutions of the more difficult partial differential equations describing wave motion in two or three dimensions. Moreover, he shows an awareness that the vibrating string phenomenon is only a special case of wave motion: at the end of the paper, he mentions an initial configuration of the form

The 'odd and periodic' extension of this graph gives a motion of the hump which is periodic, and yet in no sense an 'oscillation' of the whole string. Since such a curve can have no single 'analytical' representation and a fortiori no trigonometrical one, this example is crucial against Euler's adversaries. I am mildly surprised that in his extended treatment of this topic,[8] Euler did not mention that the phenomenon can be seen when one snaps a long cord, such as a clothes-line. Perhaps the different economic arrangements kept the eighteenth-century physicists from having such a broad range of experiences as their twentieth-century successors.

Concluding my discussion of Euler, I should say that I can hardly think of a better introduction to primary sources in mathematical physics than these three papers of Euler. They are models of lucid and reasoned argument, with a manageable technical component, are written in French, and are accessible in his *Opera*.

(b) BERNOULLI

Bernoulli's later arguments[9] are not so illuminating. Arguing the issue of the 'generality' of a trigonometrical expansion, he relied on speculative ideas on the passage of the finite to the infinite, and on the undeniably sinusoidal form of the motions of systems of particles. In his letter to Clairaut of 1758, he raised the next question: if the 'arbitrary function' does *not* appear in the *n*-particle system, where can it come from when $n = \infty$? And he makes it clear repeatedly, that the principle of the superposition of independent 'pendulum' oscillations is for him primarily a powerful heuristic idea for physical theory. He realized the application of this idea to optics, giving support to a wave theory of light in opposition to Newton's particle theory. Newton's 'fits of easy reflection' had a natural interpretation as wave-length, and the dependence of colour on wave-length was seen by Bernoulli. The extent to which Young was stimulated by Bernoulli's ideas remains to be explored.

(c) D'ALEMBERT

D'Alembert was rather slow in getting down to detailed arguments. In his article 'Fondamental' in the *Encyclopédie*[10] he made some telling criticisms of Bernoulli's theory as a physical hypothesis of the nature of sound. The strongest of these is that the secondary vibrations (such as those giving the octave), being *relative* to the primaries, are not truly isochronous. One can see this easily by imagining a system of two oscillating mechanisms, the first fixed to the ground and the second fixed to the moving part of the first. Hence the vibrations in the still air generated by Bernoulli's supposed secondary oscillations will not be isochronous, and the main basis of the theory is destroyed.

It was in his *Opuscules Mathématiques*[11] that d'Alembert produced his most sustained attack. I cannot try to summarize the contents of this bewildering jumble, which shows all the signs of having been sent to the printer page by page as it was written. As usual, I shall select a few of the best points in the article. First of all, he makes it clear that he is doing mathematics, and thinks that this is what Euler is, or should be, doing. For how can one argue from physical events when the only case actually realized in practice (plucking a string) is by universal consent excluded from the differential-equations theory (it involves a discontinuous tangent)?

His criticisms on the mathematical side centre mainly on the question of discontinuities. If one has a general 'pieced-together' $f(x)$, then this will have a discontinuity in its curvature at the points $x = 0, \pm l, \pm 2l, \ldots$

Now, when one constructs the solution $\frac{1}{2}[f(x+ct)+f(x-ct)]$, one finds that these discontinuities in curvature appear anywhere, not just at the fixed end-points of the string. But one cannot say that the solution 'satisfies' the wave equation (3) when at various points the equation has no meaning for the proposed solution, and moreover where the cord cannot be said to have a determinate acceleration. Euler's answer to this was two-fold: first to say that the 'error', occurring at only two points at most, is not significant; and also (more constructively) to consider the generalized solution of the equation, rather than the equation itself, as the description of the phenomena.

Euler's various constructions of the solution are subjected to criticism. I cannot show here how the construction of the solution in the general case of non-zero initial displacement *and* velocity is a matter of some delicacy—but it is. Moreover, d'Alembert showed how in subtle ways any construction presupposes that the (unknown) function is one defined by a regular law: this is implied by the use of the 'exact differential' ($df = f_x dx + f_t dt$). Arguing on Euler's own ground, he finds further reasons for restricting the class of initial configuration $f(x)$. For if the acceleration is zero at the end-points, then similarly the curvature must be zero: otherwise the basic equation would not be consistent with the boundary conditions. (A more modern restatement would be that curvature must be arbitrarily small in the 'neighbourhood' of the end-points—but the restriction remains.)

All these criticisms would have had far more effect if they had been given to Euler early in the 1750s. As it was, they were too late on two counts. First, as we have seen, Euler's own thinking had progressed to the point where these pure-mathematical criticisms would not worry him unduly. And second, the appearance of Lagrange's *Recherches sur la nature et la propagation du son* had completely transformed the situation, so far as Euler was concerned.

4. *Lagrange, and the end of the affair*

Lagrange's *Recherches*[12] is a stunning piece of work. A mathematical apparatus of overwhelming power is constructed, for the mounting of a decisive attack on a problem which had defied Bernoulli, d'Alembert and even Euler. Out of this came not only a physical theory, but also a new interpretation of the nature of the $f(x)$ of the vibrating-string problem; and then the gates were open for the exploration of a whole empire in mathematical physics. The fact that the mathematics was imprecise at critical points, that the physical theory was wrong, that the symbolism obscured even more important results, and that the success of the method rendered it obsolete at one stroke—all these do not detract from the magnitude of the achievement, but only emphasize its solidity, like brushwood caught in the treads of a bulldozer.

There is very strong evidence that at the outset Lagrange was not primarily concerned with the Vibrating String problem, but rather with the propagation of a pulse through an elastic medium. His approach to the problem was a time-honoured one: imagine n mass-points in a line, connected by springs, obtain the mathematical formalism describing their motions, and then transform the formulae by letting n tend to infinity. Euler had had a limited success with this problem in an earlier paper,[13] solving the special cases of two, three, or four particles and then 'inducing' a general result. Lagrange made it quite plain that he was not content with such an approach, and set up a very powerful and abstract algebraic machinery to handle the problem in all its generality. Having achieved a solution for the motions of n particles with the first having an initial velocity, he made some extremely fortunate simplifications, set $n = \infty$, and got an expression describing the propagation of a pulse in a continuous medium. This was a most unusual object, what we now call a δ function: zero everywhere except at one point, and infinite there.

The infinity could be taken care of by multiplying on a dx; but the interpretation of the formalism gave a novel theory of propagation. This is, that a 'wave' of velocity travels through the medium, each particle in turn jumping from zero to finite velocity, and back again to zero. With this theory he could explain such things as echoes, and also discuss the superposition of sounds. The theory was very short-lived.

But around this time he looked again at his mathematical machinery, and realized that it could be applied equally well to the equations of n particles on a string: a problem worked over by all the controvertists. It is this work which eventually appeared in the '*Recherches*'. There he exhibited for the first time the full solution of the n-particle problem for the vibrating string; and all of Bernoulli's superposed vibrations were there to be seen. But Lagrange was after bigger game, and with the previous work on pulse propagation to guide him, he simplified and transformed his formulae, getting an expression for the wave-form which we now write as $y(x, t) = \int f(u)[\frac{1}{2}\delta(x + ct - u) + \frac{1}{2}\delta(x - ct - u)]du$.

Here, for Lagrange, was the proof against Bernoulli: as the number of particles increase, the superposed waved *do* pile up in such a fashion as to be quite transformed. The mathematical illustration (and basis) of the effect is given by the difference between
$$\cos x + \cos 2x + \ldots + \cos nx,$$
a periodic function for any n, and
$$\cos x + \cos 2x + \ldots + \cos nx + \ldots$$
which is infinite for x equal to zero, and undetermined but effectively zero in an integration, for x different from zero.

There has long been a bit of mystery about this achievement of Lagrange's. For in the course of it he came close to, but missed, the Fourier series expansion of an arbitrary function, given (for an 'even' function) by the formula

$\frac{1}{2}\pi f(x) = \frac{1}{2}\int_0^\pi f(t)dt + \cos x \int_0^\pi f(t)\cos t\, dt + \cos 2x \int_0^\pi f(t)\cos 2t\, dt + \ldots$ (11)
In this present paper, he did not rearrange his symbolism in such a fashion as to reveal the formula; moreover, the problem from which he started, and the abstract character of his mathematical machinery, would not predispose him to look for such a result. In a later paper[14] he did exhibit a formula almost identical to the one above, and, I believe, recognized its implications. But by that time he was so firmly committed to the use of functions 'more general than those represented by an analytical formula' that he sought a way to explain it away, and treated it as an interpolation formula of arbitrarily great accuracy. The oddest thing about the controversy is that results identical with, or approaching, the 'Fourier' expansion (11) were being obtained through the 1750s, by Euler, d'Alembert, and others. But this was in connection with an entirely different problem: the trigonometric expansion of known periodic functions (or interpolation of unknown ones) for making approximations to be used in astronomical calculations. Clairaut was the one who realized the generality of such expansions.[15] Why he did not exploit this result is a mystery to me; he was friendly with Bernoulli (the 1758 paper by the latter was in the form of a letter to him), and a bitter enemy of d'Alembert.

This unexploited work on trigonometric expansions does lend support to the modern judgement that this was far from the central issue in the controversy. (It is for this reason that I have omitted the discussions of them from this survey.) None the less, the publication of such a result could not have been without effect in the controversy. For in the absence of universally acceptable proofs, arguments turned on methodological principles, and on striking evidence for one side or the other.

It was as striking evidence that Lagrange's paper had the greatest effect. For he had been able to get the functional solution for the vibrating string problem without 'doing calculus'. Hence, the problem of the implicit assumptions on the nature of the original configuration $f(x)$ (made explicit by d'Alembert in the *Opuscules* but recognized in advance by Euler) does not arise. Thus, for Euler, the derivation represented a vindication of his view that the $f(x)$ should be absolutely general. It is almost certain that Lagrange's paper was for him a liberation from his previous doubts and hesitations. Within a few months a stream of papers making essential use of the 'completely general function' idea, as in the propagation of pulses, were read by him. Moreover, he lost no opportunity to praise Lagrange's achievement, publicly and privately. It is only fair to say that there was probably a bit of politics in this; it is known that Euler had held back the publication of a controversial paper (probably *Eclaircissemens*) at the request of Maupertuis, to avoid a public wrangle with d'Alembert. Now, Lagrange was a brilliant young mathematician of great fame, and by co-opting him as an ally Euler gained a certain immunity from attack. There is evidence that Lagrange was a somewhat reluctant ally, and he enjoyed jokes at Euler's expense in later letters to d'Alembert, where he took Euler's side of the 'generality' argument.

Lagrange's position on the ideological continuum between mathematics and physics seems to be the same as that of Euler. At the beginning of his paper he comments that Newton's derivation of the velocity of sound was satisfactory to the physicists, but not to the mathematicians: and the discovery of serious errors in Newton's proof led him to investigate the problem. He is interested in the experimental results, but does not allow himself to be ruled by them. He confesses that the transmission, without mixing, of different sounds by the air is a problem which he cannot resolve, but is contented with d'Alembert's criticisms of Bernoulli's theory. I think it revealing that of this problem he should say that '*les plus habiles* (of the physicists) *ont été obligés de recourir à des systèmes pour en rendre raison*'. The connotations of '*hypotheses non fingo*' carried by these words sets Lagrange more than a certain distance towards 'mathematical' physics.

For all practical purposes, Lagrange's paper closed the controversy. He and Euler set up partial differential equations, and derived 'functional' solutions, in a variety of problems in continuum mechanics, and it was this work which led on to the achievements of the nineteenth century in this field. Although the controversy spluttered on, nothing significant came out of it. Finally, the work of Monge in the 1770s,[16] giving a geometrical interpretation to the integration of partial differential equations, seemed to provide a conclusive proof of the fact that functions 'more general than those expressed by an equation' were legitimate mathematical objects, in this field at least.

The greatest virtuosity in the use of unknown 'arbitrary functions' in the solution of difficult hydrodynamical problems was shown by Poisson, in his paper on sound of 1807.[17] He had every justification for summarizing the outcome of the vibrating string controversy in the following words:

A l'époque où elles (Lagrange's results) ont été publiées, le calcul aux différences partielles, d'où depend la solution de ce genre de questions, était à peine connu; on n'était pas d'accord sur l'usage des fonctions discontinues, qu'il est cependant indispensable d'employer pour représenter l'état de 'lair au origine du mouvement; heureusement les progres de l'analyse ont fait disparaitre ces difficultés, et celles qui subsistent encore tiennent à la nature de la question.

At just about the same time,[18] Fourier gave a brief review of the vibrating string controversy. He quoted d'Alembert's point that an infinite sum of sine curves must have a continuous curvature (thereby rendering Bernoulli's formalism inapplicable to the physically realized case of the plucked string), and Euler's use of a curve different from zero for only a part of its interval of definition (thereby excluding Bernoulli's formalism from the more general problem of pulse propagation). Then he continues:

Ces objections font assez connaître combien il était nécessaire de démontrer qu'une fonction quelconque peut toujours être développée en series de sinus ou de co-sinus d'arcs multiples; et de toutes les preuves de cette proposition, la plus complette est celle qui consiste à résoudre effectivement une fonction arbitraire en un telle série, en assignant les valeurs des coefficens.

And this Fourier has just done; and he has shown that the coefficients can be calculated for functions having different definitions over different parts of the interval, having discontinuous tangents, and even having jump-discontinuities. Thus he can say:

Les théorèmes précédens satisfont à cette condition, et je me suis convaincu, en effet, que le movement de la corde sonore est aussi exactement représenté dans tous les cas possibles, par les developpemens trignométriques que par l'intégrale qui contient les fonctions arbitraires.

We now know that such a set of bare 'existence' theorems would not have been as 'crucial' in the Vibrating-String controversy as Fourier thought. But since in his own problem (heat conduction) he went on to demonstrate the great power of trigonometric and related expansions for mathematical physics, his was the final verdict.

NOTES

1. J. D'ALEMBERT, 'Recherches sur la courbe qui forme une corde tenduë mise en vibration', and, 'Suite des recherches...' *Berlin Mémoires, 3,* 1747, pp. 215–19.
2. L. EULER, 'Sur la vibration des cordes', 1748, E 140, pp. 63–77.
3. L. EULER, 'De motu corporum flexibilium', 1744–6 and 1751, E 165 and E 174, pp. 165–232.
4. J. D'ALEMBERT, 'Addition au mémoire sur la courbe qui forme une corde tenduë, mise en vibration', *Berlin Mémoires, 6,* 1750, 355-60.
5. DANIEL BERNOULLI, 'Réflexions et eclaircissemens sur les nouvelles vibrations des cordes exposées dans les Mémoires de l'Academie de 1747 and 1748', *Berlin Mémoires,* 1753, pp. 147–72. 'Sur le mêlange de plusieurs especes de vibrations simples isochrones, qui peuvent coexister dans une même système de corps,' ibid, pp. 173–95.
6. L. EULER, 'Remarques sur les mémoires precedens de M. Bernoulli,' 1753 E 213, pp. 233–54.
7. L. EULER, 'Eclaircissemens sur le mouvement des cordes vibrantes,' 1762–5, E 317, pp. 377–96.
8. L. EULER, 'Sur le mouvement d'une corde, qui au commencement n'a été ebranlée que dans une partie,' 1765, E 339, pp. 426–50.
9. DANIEL BERNOULLI, 'Lettre de Daniel Bernoulli, de l'Académie Royale de Sciences, à M. Clairaut de la même Académie, au sujet des nouvelles découvertes faites sur les vibrations des Cordes tendues,' Journal des Sçavans (Amsterdam), 34, 1758, pp. 59–80.
10. J. D'ALEMBERT, article 'Fondamental,' in Encyclopédie, 3rd edition, 1773, Volume 7, pp. 52–3. (The content of the article fixes a date of composition of the late 1750s.)
11. J. D'ALEMBERT, 'Opuscules Mathématiques,' Tome I, Paris, 1761, pp. 1–73.
12. J. LAGRANGE, 'Recherches sur la nature et la propagation du son,' 1759; Oeuvres, I (ed. J.-A. Serret), Paris, 1867, pp. 39–148.

13. L. EULER, 'De propagatione pulsuum per medium elasticum,' 1747–8, E 136, pp. 98–131.
14. J. LAGRANGE, 'Solution de différents problèmes de calcul intégral,' 1762–5; Oeuvres, I, 471–668; especially pp. 551–4.
15. A. CLAIRAUT, 'Mémoire sur l'orbite apparente du soleil,' *Mémoires de l'Academie Royale des Sciences* (*Paris*), 1754, pp. 521–64; especially pp. 545–9.
16. G. MONGE, bibliography in R. Taton, *l'Oeuvre scientifique de Monge*, Paris, 1951, pp. 377–8.
17. S. POISSON, 'Mémoire sur la théorie du Son,' *Journal de l'École Polytechnique*, 14ᵉ cahier, tom. 7, 1807, pp. 319–92.
18. J. FOURIER, 'Sur la propagation de la Chaleur' (unpublished memoir read to the Institute 21 December, 1807), p. 115; MS. 1851, Bibliothèque de l'École des Ponts et Chaussées. (This MS. has been microfilmed through the generous help of M. Lehanneur, Conservateur de la Bibliothèque, and the microfilm is at the University Library, Leeds.)

9

The Controversy on Freedom in Science in the Nineteenth Century

JOHN R. BAKER

THE future historian of science in England will look back on the 1930s and '40s as a period of conflict in thought about freedom in science. From 1931 onwards for several years, those who believed in the central planning of scientific research dominated the field, because there was no systematic body of thought that opposed their opinions. The need for freedom in science had so much been taken for granted for several decades past, that no one had troubled to arm himself with a consistent theory embodying the reasons why freedom was desirable in this field. It was not until 1939 that active steps were taken by anyone to oppose the current propaganda on the necessity for central planning. In that year Professor Michael Polanyi published his paper on *Rights and Duties of Science*.[1] The story of the reaction among scientists against the central planners has been told elsewhere[2] and need not be repeated here. For the present purpose it is only necessary to note that throughout the long controversy, no mention was ever made of the discussion of a similar topic in the previous century. Yet *die Freiheit der Wissenschaft* had been a familiar phrase to many, well over half a century before Professor Polanyi wrote his article.

The oblivion into which the earlier controversy had fallen is all the more inexplicable in view of the celebrity—one might almost say notoriety—of the two protagonists. It is surprising that a head-on clash between such contestants as Haeckel and Virchow should be forgotten. The encounter was memorable in more ways than one—not least because the two men might have been expected to take their places on the same side of the battlefield. For these two masterful personalities of the biological world had much in common.

Virchow was by thirteen years the senior. As a young doctor he was deeply shocked when he saw for himself, in an epidemic of typhus, how poverty could generate disease. He turned towards liberalism, and his political convictions never changed from that time onwards. Throughout life he was able to combine two deep interests—in science on one hand (especially cytology, pathology, and anthropology), and in politics on the other. He was a dominating person: not only a thinker, but a doer; restless in his scientific and public activities; vehement and passionate, too, even in some of his scientific writings. He reached his highest achievement in the *Cellularpathologie* of 1859,[3] in which he firmly established pathology on a cytological basis and expounded the origin of cells from pre-existing cells, in endless series.

Haeckel's education had been in medicine, like that of so many biologists of his time, and his teacher in pathology at Würzburg was no other than Virchow himself. Like the latter, he was both scientist and politician, and he, too, was a liberal. In the Germany of his day the 'advanced' thinkers in politics and religion were the chief adherents to Darwin's theory of the evolution of living organisms in general and of man in particular. And it was here that Haeckel found his bent: for although he was no mean investigator, yet it did not fall to his lot to make many striking discoveries of a factual kind. His fame rested chiefly on his skill in expounding the doctrine of evolution. In his love of speculation he seemed almost like a re-incarnated Oken. Like Virchow, he was a doer as well as a thinker, and he set out to influence public opinion on man's place in nature; but he was so carried away by enthusiasm that some of his writings—especially *Die Perigenesis der Plastidule*[4]—were little short of fantastic.

The great debate between these two men took place at the Congress of German Naturalists and Physicians held at Munich in September 1877.[5] Although eighteen years had elapsed since the publication of *The Origin of Species*,[6] Haeckel felt it necessary to use the occasion to support the theory of evolution. He claimed that exact or experimental proofs were only applicable to a part of science. Elsewhere, and especially in morphology, it was proper to adopt the 'historical–philosophical' method, since organisms could only be understood through a study of their evolutionary history. He admitted that the theory of recapitulation was not capable of direct experimental proof, yet he claimed that its validity was similar to that of the accepted theories in history, archaeology, and linguistics.

Haeckel let himself go on a favourite theme—the 'soul' in animate and inanimate nature. The single-celled animals showed sensitiveness, imagination (*Vorstellung*), volition, movement. The 'Monera' or still simpler organisms were also sensitive and capable of movement. This was only possible if the *Plastidule* or molecules of protoplasm were themselves of similar nature, and it was therefore necessary to believe in the existence of a *Plastidulseele*. Since the combination of carbon with hydrogen, oxygen, and sulphur to form *Plastidule* involves the production of a soul, it was legitimate to speak also of atom-souls. Haeckel considered these beliefs to be consistent with his monistic philosophy, which, in his view, bound together practical idealism and theoretical realism—whatever that may have meant.

He now turned to education, the main theme of his address. He opposed the teaching of mere facts to the young. Both teacher and pupil would approach their work with infinitely more interest if they always asked themselves, '*Wie ist das enstanden?*' Causes, not simply their results, must be the subject of study in the schools. He demanded the teaching of the evolutionary doctrine in the fields of cosmogeny, geology, biology, anthropology, and linguistics.

Haeckel went on to a direct attack on 'church religions', with their ancient myths about creation, and urged that they should be supplanted by a *Naturreligion* based on the 'natural moral law' that evolved from the social instincts of animals. The future, he claimed, did not belong to the kind of theology that wages a fruitless war against the victorious doctrine of evolution, but to a religion that takes possession of the doctrine, recognizes it, and utilizes it.

Virchow had recently made known his attitude towards the doctrine of evolution.[7] He had deplored the tendency to unwarranted generalization exhibited by the Darwinists, whom he compared with the *Naturphilosophen* of Oken's school. He had emphasized the danger of wild hypotheses and the need for demonstrable facts. He himself had been prepared to accept the mutability of species even before the publication of the *Origin*, but he felt that the younger generation needed to be recalled to prudence.

Virchow had not arrived in Munich when Haeckel gave his address, but he obtained a copy of it. He saw at once how Haeckel's whole outlook conflicted with his own. He considered that a further warning was necessary. It appears that he had intended to address the Congress on another subject, but decided to change his plans and issue a direct challenge to Haeckel. Four days after the latter had spoken he made a remarkable extempory speech, which was subsequently published.[8] He must have been in genial mood, for his words were interrupted no fewer than nineteen times by the merriment and approval of his audience; but his purpose was serious enough. As a deputy to the Prussian parliament, he was anxious lest the nation should lose confidence in its scientists, and this would happen if they abandoned themselves to speculation. Science would suffer if theories were expounded to the public and subsequently found to be untrue. If the doctrine of evolution could be proved, it should be taught in the schools, even though it were bound up with socialistic opinions; but he emphasized how fragmentary human knowledge was ('... *alles menschliche Wissen Stückwerk ist*') and pleaded for humility ('*Das, was mich ziert, ist eben* die Kenntniss meiner Unwissenheit').

He attacked Haeckel's claim that all ideas concerning the spiritual life should be based on the *Plastidulseele*. Until one could define the properties of carbon, hydrogen, oxygen, and nitrogen in such a way as to show that a soul must necessarily arise from their combination, there should be no teaching on the subject in the schools.

In a remarkable and unexpected passage, reminiscent of Polanyi's *Science, Faith, and Society*,[9] Virchow admitted that faith has its place even in science ('*Es gibt in der That auch in der Wissenschaft ein gewisses Gebiet des Glaubens*'). In both science and religion there were three parts— a certain quantum of objective truth, a broad middle path of faith, and a region of subjective and fanciful ideas. He himself could not entirely renounce the subjective spirit in science, but he saw how necessary it was to be objective.

He stated clearly the necessity for freedom of inquiry (*die Freiheit der Forschung*), and recognized that research-workers must pose and discuss their *Probleme*; but these must not be taught to the public as doctrines. He himself would not be surprised if it were found that man had animal ancestors, but there was no proof yet. Intermediates might be found in the future, but the known fossil men were not more ape-like than the most primitive people living at the present time.

Virchow ended by assuring the audience that scientists would receive from the government all necessary security and support in whatever directions their *Probleme* might lead them. His speech made a profound impression, not only on those present, but throughout Germany.

It is to be noted that Virchow drew a clear distinction between *die Freiheit der Forschung* and *die Freiheit der wissenschaftlichen Lehre*. The former he demanded as a necessity; but he feared the teaching of speculative ideas in schools, because he thought that the reputation of science would be endangered.

Haeckel had left Munich before Virchow arrived, but the latter's speech was already available in print next month. Haeckel was not the man to leave such criticism unanswered. His reply took the form of a small book, first published in the year after the Munich conference,[10] and subsequently translated into English and provided with a preface by T. H. Huxley.[8] This work is less fanciful than many of Haeckel's writings. It gives the impression of careful compositon. The argument is effective, despite the polemical style.

Haeckel's main purpose appears to have been to discredit Virchow by a personal attack. He pays tribute to all that he owes to his former teacher, and even allows that it was Virchow who first converted him to monism, the philosophy that dominated the rest of his life. He claims, however, that Virchow has undergone a profound change in outlook. The latter was formerly an inspiring teacher just because he was willing to discuss his ideas with his students, instead of confining himself to the recitation of demonstrable facts; he did not by any means reject the publication of hypotheses in his earlier days. His famous generalization, *Omnis cellula e cellula*, was only a theory, not a fact; indeed, it was not universally valid, though it contained an important element of truth.[12] Now, after the lapse of years, Virchow claims that such theories should not be taught to students.

With some justice, Haeckel lays emphasis on Virchow's ignorance of morphology, especially so far as the lower animals are concerned, and does not hesitate to agree with the latter's confession of the limitations of his own knowledge.

Haeckel ridicules the idea that nothing is known except what can be proved by experiment. Virchow allows that fossils actually represent the remains of extinct animals, though no absolute proof of this can be brought forward, and no experiment devised to settle the matter. The Mesozoic mammalia are known only by their lower jaws. It is reasonable to assume that the rest of the skeleton was mammalian. Virchow, however, to be consistent, must suppose that in these remarkable creatures the only skeletal element was the lower jaw!

Haeckel uses his full capacity as a polemical writer to attack Virchow's suggestion that the doctrine of evolution has some connection with socialism. Virchow's vague remarks on this subject are interpreted by Haeckel as meaning that the horrors of the Paris Commune of 1871 were somehow to be regarded as a consequence of the spread of Darwin's ideas. Haeckel points out with some effect that the reverse is true; that Darwinism could be used, on the contrary, to support aristocratic régimes. Socialists, he says, should try to stifle the doctrine, for its teaches that inequality is essential to progress. He makes it clear that he is no socialist himself, and remarks savagely on '*den bodenlosen Widersinn der socialistischen Gleichmacherei*'.

He expresses his surprise that Virchow should require scientists to limit their teaching to demonstrable facts, while clerics are permitted to instruct their pupils on miracles and religious dogmas.

Here and there Haeckel descends to downright abuse. For instance, he describes Bastian (a supporter of Virchow) as the 'Acting Head Privy-Councillor of Confusion'. He blames Virchow for being too much bound up in political life, for arguing like a Jesuit, for wasting his time in measuring human skulls, and for moving from a small university (Würzburg) to a large one (Berlin). He views with horror the possibility that German science might be centrally planned under the domination of the capital city, and castigates the biologists of Berlin University for their misunderstanding and disregard of the theory of evolution.

Huxley's preface to the English edition of Haeckel's book is a sparkling little essay in characteristic vein.[13] One might perhaps have expected him to come down heavily on Haeckel's side, for the two men were personal friends and both were public champions of the evolutionary doctrine. He gives full credit, however, to Virchow, by whose arguments he is clearly impressed. Indeed, he finds himself able to agree with every important general proposition in Virchow's address, and is surprised that it should have been so unfavourably received by some.

Huxley realized that the advance of science required the collaboration of men of two very different kinds. 'The one intellect is imaginative and synthetic; its chief aim is to arrive at a broad and coherent conception of the relations of phenomena; the other is positive, critical, analytic, and sets the highest value upon the exact determination and statement of the phenomena themselves.' Huxley mentions that neither Haeckel nor Virchow is a typical example of either of these schools. He might have said that Haeckel was an extremist of the first, and that Virchow himself could lay little claim to represent the second, however much he might admire it.[14]

Although Huxley is considerate towards Virchow, he rebukes him for trying to establish a connection between the evolutionary doctrine on one hand, and socialism and revolution on the other.

In his book, Haeckel had made the sensible observation that it was for the teachers to decide what should be taught. There is, indeed, no freedom of learning if a central organization determined the curricula of all the educational establishments in a country. It is essential that some at least of the schools and universities should be autonomous. Article 20 of the Prussian Charter laid it down that '*Die Wissenschaft und ihre Lehre ist frei*', but in fact there was central planning of all education under a Minister. In a sense the latter supported Virchow, for he forbade schoolmasters to teach Darwinism; but at the same time he excluded biology from the curricula of all senior classes in schools.[15] This was a result of the controversy that neither Haeckel nor Virchow could have foreseen and both must have deplored.

This account of the controversy on freedom in science in the 'seventies of the last century will have shown how radically it differed from the more serious and more protracted struggle of the 1930s and '40s. In the latter, the whole of science was at stake. If the propaganda for the central planning of all scientific research had succeeded, science could scarcely have survived. In the 'seventies no one had any intention of restricting the freedom of research workers to choose the subjects of their investigations. The proposed restriction was in a less important field, and it is not possible now to assert exactly what restriction Virchow proposed. Did he really intend that lecturers at German universities should not be permitted to discuss scientific theories with their students? It seems scarcely credible, and in fact he did not make any clear statement to this effect. It seems that he was thinking rather of schools, where it is appropriate that the instruction in science should be mainly factual. He was genuinely alarmed. He feared that under Haeckel's influence there might be a recrudescence of the *Naturphilosophie* of Oken. He himself had come under Oken's sway in early life, had seen the errors of *Naturphilosophie*, had revolted against it, and now showed a convert's enthusiasm.

It is very doubtful whether this controversy would have flared up as it did, if it had not been concerned with the origin of man. It is curious that neither of the contestants gave any serious consideration to the direct fossil evidence on this subject. Indeed, Haeckel stated specifically that fossils showing characters intermediate between those of ape and man were not to be expected,[16] and Virchow remarked that no fossil man was more ape-like than certain existing men.[17] Yet Neanderthal man, who had been known for many years, does show ape-like characters in many parts of his skeleton. Although nowadays he is not thought to be on the direct line of ancestry of modern man, yet it is surprising that the early Darwinians paid so little attention to him. Darwin himself dismissed him in a sentence, merely remarking on his large cranial capacity.[18] Huxley, however, in his preface to the English edition of Haeckel's book,[19] denies (in a footnote) Virchow's claim that no intermediate fossil is known. He says that no modern human cranium is so pithecoid as that of Neanderthal man. Although this is true, yet fossil evidence of man's ancestry was scanty in the 1870s, and Virchow's attitude was justifiable.

NOTES

1. M. POLANYI, 'Rights and Duties of Science,' *Manch. School Econ. Soc. Stud.*, Oct., 1939, p. 175.
2. ANON., *The Society for Freedom in Science: its Origin, Objects, and Constitution.* 3rd edition, Oxford, 1953 (published by the Society).
3. R. VIRCHOW, *Die Cellularpathologie in ihrer Begründung auf physiologischer und pathologischer Gewebelehre.* Berlin, Hirschwald, 1859.
4. E. HAECKEL, *Die Perigenesis der Plastidule oder die Wellenzeugung der Lebensteilchen*, Berlin, Reimer, 1876.
5. E. HAECKEL, 'Ueber die heutige Entwickelungslehre im Verhältnisse zur Gesamtwissenschaft,' *Versamm. deut. Naturf. Aerzte*, 50, 1877, pp. 14–22. R. Virchow, *Versamm. deut. Naturf. Aerzte*, 50, 1877, pp. 65–78 (no title).
6. C. DARWIN, *On the Origin of Species by Means of Natural Selection*, London, Murray, 1859.
7. R. VIRCHOW, 'Ueber die Standpunkte in der wissenschaftlichen Medicin,' *Arch. Path. Anat. Physiol. klin. Med.*, 70, 1877, pp. 1–10.
8. R. VIRCHOW, *Versamm. deut. Naturf. Aerzte*, 50, 1877, pp. 65–78; *Die Freiheit der Wissenschaft im modernen Staat* (reprint of previous title), Berlin, Wiegandt, 1877; *The Freedom of Science in the Modern State*, 2nd edition, translated with a new preface by the author, London, Murray, 1878.
9. M. POLANYI, *Science, Faith, and Society*, London, Oxford University Press, 1946.
10. E. HAECKEL, *Freie Wissenschaft und freie Lehre. Eine Entgegnung auf Rudulf Virchow's Müncher Rede uber 'Die Freiheit der Wissenschaft im modernen Staat'*, Stuttgart, Schweizerbart, 1878.
11. E. HAECKEL, *Freedom in Science and Teaching* (translated). With a prefatory note by T. H. Huxley, London, Kegan Paul, 1879.
12. R. VIRCHOW, *Die Cellularpathologie* (see note 3 above).
13. E. HAECKEL, *Freedom in Science* (see note 11 above).
14. It may be mentioned in passing that Virchow has been too much praised

for his contributions to the cell-theory. Remak (1852, 1855), not Virchow, made the first clear and solidly-backed general statement of the way in which cells multiply. The coining of the catch-phrase, *Omnis cellula a cellula* (1885) or *e cellula* (1859), gave Virchow an underserved advantage over Remak. See J. R. Baker, 'The Cell-theory: a Restatement, History and Critique. Part IV. The Multiplication of Cells.' *Quart. J. Micr. Sci.*, 94, pp. 407–40. Cf. F. Remak, 'Ueber extracellulare Entstehung thierischer Zellen und uber Vermehrung derselben durch Theilung,' *Arch. Anat. Physiol. Wiss. Med.*, 1852 (no vol. no.), pp. 47–62 and *Untersuchungen uber die Entwickelung der Wirbelthiere*, Berlin, Reimer, 1855.

15. E. NORDENSKIÖLD, *The History of Biology: A Survey* (translated), New York, Tudor, 1946, p. 522.
16. E. HAECKEL, 'Ueber die heutige Entwickelungslehre' (see note 5 above).
17. R. VIRCHOW, *Die Freiheit der Wissenschaft* (see note 8 above).
18. C. DARWIN, *The Descent of Man and Selection in Relation to Sex*. Second edition, London, Murray, 1874.
19. E. HAECKEL, *Freedom in Science* (see note 3 above).

PART THREE

The Knowledge of Society

The laws and the morality of a society compel its members to live within their framework. A society which accepts this position in relation to thought is committed as a whole to the standards by which thought is currently accepted in it as valid. My analysis of commitment is itself a profession of faith addressed to such a society by one of its members, who wishes to safeguard its continued existence, by making it realize and resolutely sustain its own commitment, with all its hopes and infinite hazards. *P.K.*, p. 321

10

Max Weber and Michael Polanyi*

RAYMOND ARON

THE reader will be surprised by the conjunction of these two names. The first does not occur in the index of *Personal Knowledge* and there is nothing to suggest that Polanyi has studied the work of the German sociologist. The intellectual temper of the two men appears as different as possible. The first is a philosopher of contradiction dedicated to science, but in suffering, with the covert sorrow of being excluded by the progress of science from the paradise of faith. The second is a philosopher of reconciliation, convinced that it is only through a misunderstanding of its true nature that science disenchants the universe. Between the knowledge of the verifiable and the intuition of the inexpressible, the former establishes a radical break and the latter a continuous progression.

Nor indeed is there any question, in the following pages, of comparing the minds or works of the author of *Wirtschaft und Gesellschaft* and the recipient of this volume. I was tempted, to begin with, to apply the theory of *Personal Knowledge* to the social sciences in general and to sociology in particular. But since this project seemed to me, on reflection, to exceed the scope of a short essay, I thought that the analysis of an *objectivist*† conception of social science would constitute an introduction to a project of this kind.

It might be denied that Max Weber developed an *objectivist* theory of knowledge; that, on the contrary, he bound knowledge to the person of the knower, since the constitution of the object and the selection of the facts stem from an extra-scientific choice which may fairly be called *personal*. I should not dream of denying that Max Weber, admitting an

---

* When I wrote this essay, I had not read Professor Polanyi's *The Study of Man*, in which some of the problems touched on in the following pages are explicitly treated. On reflection, I have let the text stand, except for some minor corrections on the last few pages. My intention was to take my inspiration from the ideas of *Personal Knowledge*, not to expound them. In relation to *The Study of Man*, the reader may judge to what extent my analysis agrees or disagrees with Polanyi's thought.

† In the sense given to that term by Michael Polanyi.

objectivist interpretation of the natural sciences, sought to mark out the specific traits of historical or cultural science, traits inseparable from the human character of the field to be explored. But social science seemed to him *not* to be scientific, precisely to the extent to which it was personal. Thus he forced himself to separate, within knowledge, the universally valid parts from the subjective and historical elements, in which, indeed, he did not deny the inevitable intervention of the knower, but which seemed to him contrary to the essence of scientific research.

In other words, the epistemology of Max Weber represents a supreme effort to take account of the social sciences, and to establish and limit their objectivity, within the framework of a critical philosophy. Starting from the difficulties to which this philosophy was driven, I shall ask what contribution the post-critical philosophy of Polanyi might make to the theory of sociological knowledge.

1

Max Weber, as we know, borrowed his conceptual instruments from the so-called South-West German school, and in particular from Rickert's famous book *Die Grenzen der naturwissenschaftlichen Begriffsbildung*. He retained its point of departure, the inexhaustible and unformed given, and the two possible orientations of conceptualization, that is, either the generalization of the natural sciences or the unique singularity (einmalige Einzigartigkeit) of history. In this latter direction a second division arose: the reference to values permits us to grasp the individual characters of the elements, to specify the subject-matter as cultural, but the sociologist is no less justified in establishing general propositions relative to cultural phenomena as well as in tracing the unique features of this or that cultural unity. The fundamental distinction, therefore, is that of the sciences of nature and the sciences of culture, the former establishing their subject matter by the analysis of constancies or regularities, the latter in relation to values, and both of them including, though in very different proportions, both generalizations and propositions relative to unique facts.

The criticism of Rickert furnished Max Weber with a language and gave him, as it were, a clear conscience. History, sociology, political economy could be sciences and universally valid, without sacrificing their own distinctive appeal, the originality of their subject matter, and their desire to grasp what is original in that subject matter: that is, to know the human character. It is from this character, according to Max Weber, that the intelligibility of historical and social facts proceeds. The facts are comprehensible (*verständlich*), we can interpret (*deuten*) them directly. On this point, Max Weber, it seems to me, is following Jaspers, or more precisely Jaspers's conception of psychopathology.* Human behaviour remains comprehensible in its psychological texture up to the

* Expounded in his earliest book, *Allgemeine Psychopathologie*.

moment when it becomes pure reflex or complete alienation, so that the logic of neuroses and in part that of psychoses can be understood. Max Weber does not deduce the meaning of human behaviour from its relation to values, but sees in that relation the self-evident specificity of such behaviour (and, to a much lesser extent, of the behaviour of animals).

Since it is the nature of social science to *understand* (it is *verstchende Wissenschaft*), the crucial problem turns out to be that of discriminating between various interpretations (*Deutungen*). Most interpretations are in fact intrinsically probable, since they establish relations between motives and acts, between drives and reactions, between environment and decision, consonant with our psychological knowledge or with intuitions which though prescientific are beyond the range of doubt (that the victim of humiliation desires revenge or that the general desires victory needs no demonstration). Hence the key problem of the sciences of man, which can always explain but find their explanations hard to verify: how can we distinguish between the *probable* and the *true*, or between an intrinsically intelligible account and what really took place?

To this question, Max Weber has given a series of partial answers, which are convincing in themselves but fail to solve the essential uncertainty.

These partial and several answers are scattered throughout the expositions of the *Wissenschaftslehre*. They concern the precautions that must be taken in order not to confuse the intelligible account as such (the propositions relative to the equilibrium of the market or the logic of a certain legal system) with the immediately experienced reality to which, in approximation, these propositions or this logic may apply. *Real* capitalism must not be confused with the ideal type of capitalism which we are able to construct; the ideal type of market economy is not realized in historical market economies; the law conceived by a judge, a barrister or a litigant differs from the law elaborated by a theoretician according to what he believes to be the necessary implications of the texts. And the sociologist aims at the experienced meanings, that is, at capitalism, market economics, and legislations as they have been or as they have been experienced. Only the subjective sense is meant, because only it is or has been real.

Thus at one blow Weber's 'ideal types', so christened because they do not belong to the species and genera of botanical or zoological classification, or to Aristotelian logic, become mere instruments. Resorting to the principles of nominalism, Max Weber admits, or rather he proclaims, the right of the sociologist or the historian to remodel the object of his study by remodelling the questions he puts to it, that is to say, the concepts he applies. Each person, depending on the interest that inspires him, will construct a different ideal type of capitalism, and each of these ideal types will underline one of the aspects of real capitalism, an aspect chosen by the question that the sociologist has asked. Thus the subjectivity of the knower, itself historically determined, expresses itself in the inquiry; but this relativity of inquiry does not communicate itself to the science, since the latter consists of non-subjective replies to subjective questions, of verified answers to questions freely put.

This theory of knowledge would be satisfactory on either of two conditions: (1) if it were possible to distinguish in reality between question and answer, or (2) if the experienced meaning could be rigorously grasped. But neither of these two conditions is satisfied, or at least, Max Weber has not demonstrated that they are effectively satisfied. The discrimination between the experienced (real) meaning and the meanings produced by the rationalizations of the interpreter, and the verification of the agreement between experienced meaning and interpretation, which should have been the centre of a theory of understanding, appear only here and there in the course of his arguments. Moreover, Max Weber placed in the category of ideal type all or nearly all the concepts of the social sciences, whether the elaboration of an individual whole (Puritan capitalism) or an historical concept with several examples (western city), or the rational relations of economic theory, and hence he was unable to recognize the diversity of the relations between ideal types and reality, the diversity of senses in which one can speak of experienced meaning and the diversity of ways in which interpretations may be verified or confirmed.

Max Weber himself oscillated between the *objectivism* of the experienced meaning and the *historicism* of free choices. It is not sufficient, he said, that the interpretation should be satisfying for the mind; it must be true. That is, it must have been realized in the event or it must be verified by the constancy of its regular occurrence. But on the other hand, humanity, in its unpredictable development, ceaselessly creates new values and interrogates its past in the light of the values which arise out of the present. To forbid the living to interrogate the dead with reference to the values experienced today would be to forbid the creation of new values, or to decree that the dead have no more to say to us. The dialogue of the present and the past confers on historical science a *personal* character (in Polanyi's sense). But this character is hard to assimilate to the critical conception of science.

Max Weber borrows from the Kantian tradition the radical opposition between facts and values, between the *is* and the *ought*. There is no common denominator between ascertaining what, in fact, happened once or is happening, and determining what should be the conduct of the statesman or the citizen. But Weber goes beyond that classic antithesis. In order to save freedom as he understands it, he asserts that each person's choice of his values is arbitrary, essentially irrational, without any possible differentiation between reasonable and unreasonable choice. He invites each of his hearers to follow the voice of his genius—whether 'god or devil'. Moreover, the choice is essentially polemical, for values are not only numerous but contradictory. A thing may be beautiful just because it is not moral. Not to resist evil is cowardice, unless it is holiness. The man who chooses holiness renounces the defence of his country.

Free in his choice in consequence of the plurality and incoherence of values, man is free in his conduct also because historical reality does not admit global determinism. Puritanism is not *the* cause of capitalism as such; it has been, in fact, one of the conditions for the emergence in the west of certain specific traits of capitalism. Every causal relation, whether accidental or adequate, is partial and analytic; it results from cutting up reality into fragments; it does not apply to the sequence of totalities. But this process of dissection is inseparable from our way of asking questions and hence from our concepts. Causal relations are not arbitrary, but they are dependent on an arbitrary choice of values or of questions.

At this point, we may well ask if Max Weber has not finally dissolved the universal truth of science which he so passionately desired to secure. To begin with, scientific truth itself, in his philosophy, was only one value among others, and its affirmation was neither more scientific nor more rational than its negation. Besides, when Max Weber demanded imperiously of his hearers or his adversaries that they distinguish as he did between facts and values, science and politics, he was inviting them to subscribe to his philosophy and to renounce their own. Why should not the revolutionary adorn his own convictions with the prestige of science if there is no more merit in being reasonable than impassioned, if success is the supreme objective in politics?

Not only does science cease to be justifiable by reason as soon as it becomes the object of an arbitrary choice, but the role of universal truth in a science founded on many and contradictory questions remains indeterminable and evanescent. True, the social scientist aims at the real, and the latter does not permit just any interpretation he likes. But neither history nor sociology is able to discover so many causal connections that one can safely abandon to subjectivity all that would not be a neccesary succession, unique or regular. Besides, social science is interpretation of meaning; and all interpretation has its place within a particular conceptual system. Is the plurality of conceptual systems legitimate or fruitful? Or does it rather mark the defeat of a science claiming universal validity? If the plurality of interpretations is legitimate, must we not reconsider the alleged irrationality of the choice of values, the incoherence of the questions asked? Must we not find, at a higher level, an ordering of questions, such that their multiplicity would then cease to be random?

In short, we should have to question the two antitheses which dominate the philosophical form in which Max Weber cast his scientific experience: established or demonstrated facts as against values freely affirmed; the arbitrary character of questions put to reality as against the universal truth of the replies. The replies cannot be universally true if the questions are arbitrary; and besides, why seek with such passion for universal truth if it is worth no more than any other existential decision?

2

It is not possible here to attempt even a résumé of the principal theses of *Personal Knowledge*. I shall presuppose an acquaintance with the book and limit myself to extracting from it some propositions or maxims which seem to me to be suggested, either implicitly or explicitly, by Professor Polanyi and which might serve as introduction to a theory of historical or sociological knowledge.

(i) *Only the method of inquiry, of research or of demonstration in fact practised by the creative scientist can reveal the nature of scientific knowledge: its subsequent arrangement in hypothetico-deductive form, whatever its utility, disguises the intention of the knower and consequently the essence of his knowledge.* The same proposition might be translated into the following terms: the theory of science must not ignore the psychology of the scientist. It goes without saying that this last statement must be taken in a special sense. The psychology of the scientist that interests epistemology is that of the scientist as such, not of the man X, Y or Z, who lives in such and such a century, loves his wife and children, and is a good or a bad patriot, but of Poincaré or Einstein in search of scientific truth, of the intellectual passion which inspires them in that search, of the sense of beauty which makes them accept this or that vision of reality. In other words, the meaning of science cannot be grasped unless we start from the intention of the scientist. But this intention is not that of the *historical men* who have created science in such and such epochs or societies. It is the intention of the *scientists* who have lived in different historical contexts but have had in common the will to truth and the same conception of what a true, rational, beautiful explanation of nature must be.

Elsewhere, Polanyi affirms, 'psychology cannot distinguish by itself between true and false inferences, and hence is blind to logical principles; but it can throw light on the conditions under which the understanding and operation of correct logico-mathematical reasoning may develop, and it may supply an explanation for errors in reasoning'.* But later he points to a closer kinship between the theory of science and psychology. The latter studies the process by which living beings *learn*, it provides a theory of *learning*. And the theory of scientific knowledge is the last stage of a theory of learning which is in the first instance psychological: 'it may seem questionable whether, in the study of learning, the acknowledgement of rightness which accounts for the success of learning and accredits its achievements with universal intent, may be lumped together with the study of the conditions and shortcomings of learning. My answer is that the distinction in question is sharply pronounced only in the case of highly formalized logical operations. It becomes blurred and should be allowed to lapse altogether, when rightness is achieved according to vague maxims which

* *P.K.*, p. 337.

are effective only when applied with exceptional skill and understanding. Such, I believe, is the case for inductive inferences. The analysis of such operational principles is so closely interwoven with a study of the conditions under which they can operate or fail to operate, that the two aspects of the subject must be treated jointly. Thus in spite of the logical and epistemological affirmations contained in the theory of learning, we shall accept it wholly as a branch of psychology and authorize this branch to study—as all biology does—certain achievements ascribed to living beings'.*

In the social sciences, the desire of the theorist to separate the psychology of the knower and the logic of knowledge is so much the stronger since knower and historical person seem more indistinguishable from one another. The sociologist of today who analyses the mechanisms of the market or the administration of planning, who compares the classes of capitalist societies and the 'groups' of soviet societies, himself belongs to one or the other social type, his preferences tend this way or that. If the intention of the historian is not separated from that of the historical person, or the intention of the sociologist from that of the citizen, is not the *scientific* intention, that of universality, at once condemned? Or must we say that in the matter of societies and of colour, love and hate inspire our curiosity and become the springs of knowledge?

It was to resolve this difficulty that Max Weber was forced to effect a rigorous distinction between subjective questions and objective replies and so assumed that the knower, having expressed his historical subjectivity in the question, proceeded to efface himself before the object. It is true that the historical person becomes an historian, in the modern sense of the word, only on condition that he transcends, through the effort of knowledge, the temptation to justify, to condemn, and to distort. The error of the Weberian formulation is to transform into an objective duality a duality in fact internal to the consciousness of the knower. Thucydides, who had been an unfortunate general, admired Pericles, detested Cleon, respected Nicias; it is the man who paints these characters, not a transcendental consciousness or a pure spectator. But if the historical account is never depersonalized, neither is it ever reduced to a simple elaboration of prejudices and passions, at least so long as it preserves a scientific character in the eyes of competent men.

Historical or social science is not a sort of theology, *fides quaerens intellectum*; it is a trial, a trial of spontaneous passions in contact with facts, of prejudices confronted with the prejudices of others, a trial of the legendary past or of the contemporary scene through the effort of understanding what has happened or why it has happened. The intention of history and sociology does aim in a certain fashion at universality, but that universality is defined by the enlargement of consciousness through the criticism of itself and its institutions. The objectivist superstition, on the other hand,

* ibid., p. 370.

leads in the social sciences either to minute inquiries, purely empirical and of no ultimate significance, or to a pseudo-system of the world and of history, like that of Auguste Comte or of Marx. Either outcome is absurd.

(ii) In fact—and this maxim holds against the temptation to seek absolute objectivity through empirical moderation—it is unreasonable to sacrifice the *interesting* to the *demonstrated*. In the last analysis, no body of scientific knowledge is ever wholly and definitively demonstrated. To recognize the impossibility of demonstrating an axiom system and the rules operative within that system is not a defeat of the mind, but the recall of the mind to itself. Formalization and axiomatization are necessary, legitimate, and fruitful. But they do not create a collection of truths which would hold good all alone, on their own, without the intervention of the knower's endorsement, the commitment of the person. Still more in physics or in biology, there always exist, in every age, laws or explanations which are admitted as true without our being certain of their truth or able to demonstrate them. In short, even in the natural sciences, there are degrees in the certainty and rigour of demonstration and there are, invariably, risks of error.

In history or in sociology, if we were to be sure of escaping subjectivity and error, we should have to be satisfied with brute facts (supposing that that notion were susceptible of precise definition) or of microscopic inquiries. No historian, however positivist, no sociologist of those who have left a name, has gone very far along this path. August Comte, Durkheim, Pareto, Seignobos were no less subjective and philosophical, despite their pretensions to pure and objective science, than Scheler or Max Weber, who assigned a role to subjectivity.

Indeed, in history and in sociology, the interesting questions are also those which entail the greatest uncertainty, those which most frequently permit divergent and even contradictory interpretations. With the progress of statistics, we can succeed without too much trouble in ascertaining with precision the distribution of families in various income tax brackets. But if we ask whether there are exact divisions between these levels of taxation, or whether certain levels correspond to a particular way of living and to a 'real', in the sense of a self-conscious, 'group', our answers will be much less exact because the questions themselves utilize concepts that are not absolutely precise and are concerned with phenomena that are difficult to grasp. Nevertheless, it is not the distribution of taxes in different categories that lends interest to the problem of classes, but our interrogation about the collectivity itself: unity or plurality, homogeneity or heterogeneity, co-operation or class conflict. We must not hesitate to ask the interesting questions arising from the reality itself.

(iii) *Any science becomes self-contradictory which because of the nature it attributes to its object renders its own existence inexplicable and unintelligible.* Humanity, consciousness, science would not exist in the world envisaged by mechanists or objectivists. Rediscovering the idea of Comte and of Cournot, of a scale of 'levels of reality', the lower level being the condition or the means of realization, but not the sufficient cause of the higher level, Polanyi shows that the biologist starts from complex, living realities, and that physico-chemical explanation never accounts completely for the phenomena of life, the interpretation of which is impossible apart from reference to the finality of functions and, even before this, the distinction between the normal and the pathological.

The equivalent, in the social sciences, of the materialist attempt to explain life and consciousness by combinations of atoms or of elementary particles is the attempt of Pareto or Marx to explain the realms of the spirit by reducing them to the circumstances in which they had developed and by denying the specific intention to which each such realm owes its being. Let me explain. The interpretation of history in terms of the class struggle is presented by Marx as true; he judges it to be so. Thus he admits implicitly, in himself, his claim to universal truth. And at the same time he cannot, without contradicting himself, deny this intention in others, deny it in historical reality. The social sciences are indeed the work of historical individuals; they reflect in every period the prejudices and passions of individuals and of classes. But they would not exist as sciences if they did not also express the effort of historical individuals to test their convictions, to enlarge their knowledge, to confront their particular existence with other existences.

The analysis of the fascination of Marxism constitutes one of the most interesting chapters of *Personal Knowledge*. It is through moral passion that Marx came to deny the efficacy of morality. Because in his eyes men were untrue to the ideals that they invoked, he came, through indignation or despair, to assert that appetites and force were all that counted. He imagined a universe in which human reconciliation issued inexorably from pitiless conflict much as consciousness has been thought to issue from the collision of atoms in the universe. But the energy which animates materialists, cynics, or nihilists has still a moral origin, and this energy is increased by the sanction which pseudo-science gives to their enterprise. The latter becomes a prison for the mind as well as the body because it justifies itself at one and the same time by necessity (explicitly invoked) and by morality (unconsciously active). To escape from the prison, we must break with the first step, the denial of morality as an effective aim of humanity, and with the representation of a determinism which would assure victory in advance to one party, with whom, nevertheless, we are invited to co-operate.

Thus a sociological interpretation cannot without contradiction dissolve the scientific intention without which its own existence would be denied. It cannot without contradiction deny the moral will to which it owes its inspiration. We may generalize these remarks and assert that sociological interpretations made from outside a given universe presuppose the sociologist's understanding of the intrinsic meaning of the universe in question.

If we did not begin by understanding the painting of Tintoretto to be in some sense beautiful, the exhaustive knowledge of Venice would not suffice to reveal it to us. Without this primordial recognition of the intrinsic meanings of spiritual worlds, or if you like of culture, an avowedly mechanistic science of culture could not even begin. It must first recognize what it then pretends to reduce or to exclude.

(iv) All facts presuppose a *framework of interpretation*. The latter is never definitely demonstrated, neither deductively (since the non-contradictoriness even of a mathematical system is not demonstrable)* nor inductively, since we have never eliminated the possibility of another framework nor considered all the facts capable of refuting the framework we have provisionally accepted. In the last analysis we can say that interpretive frameworks are never either refuted or demonstrated or imposed by the facts. These propositions, classic in the epistemology of research or of scientific progress now, become integral parts of the theory of science itself. What bearing have they in the case of the social sciences?

Without trying to reproduce Polanyi's own argument on this point, I shall assert that it is just as impossible to follow through to its logical conclusion the subjectivism of Weber as it is to do the same for the alleged objectivism of Marx. The conceptual systems of the social sciences are neither the result of arbitrary choices, as Weber presents them in his theory of scientific practice, nor reflections of a structure inscribed in reality itself. Both formulae falsify what we may call the *dialogue* or *dialectic* which constitutes the authentic experience of social science and social scientists.

This thesis would need an extensive commentary. In this brief essay, I shall simply indicate the most general reasons which account for the knower's being bound by the human reality that he wishes to understand. Consider first the case of a reality where the will to express, to communicate is undeniable. The interpreter can and must first inquire, what knowledge or what wisdom the speaker wished to transmit to his contemporaries or his descendants. Philosophers, artists, scholars have wished to create, to explain; the interpreter wishes to understand their message and if he is not bound to take as final the meaning consciously given by the creator to his work, so much the less has he the right to substitute the meaning which the message would have in his own universe. If the universe of the interpreter had no kinship with the universe to be interpreted, the interpretation would be impossible. If they were confused with one another, there would be no history. Between these two extremes lies historical understanding, which entails a renewal of perspective, but implies that the historically existent subject (the knower) is aware of the otherness and uniqueness of his object, that is, of the men who are no more but whose work has been preserved.

Now all societies are, from a certain point of view, comparable to persons who have accomplished the creation of a work. A culture is a system of values and obligations which the members of a collective achieve

* It is demonstrable that this non-contradiction is not finally demonstrable.

without necessarily knowing them, as language, conceived by the linguist, is the unconscious system of spoken language.

Let us proceed to consider from this point of view a problem of social organization, for example, the problem of classes, which has been the obsession of European sociologists for so many years. The quarrels over facts and interpretations have been in large part the consequence either of the free choice of conceptual systems or of the search for a perfect objectivity. Either it was supposed that the facts examined without prejudice would say the last word on the reality of classes, or it was assumed that one had the right to decree that such and such criteria were to define a class. The second path is indeed preferable to the first. But the normal path seems to me to be intermediate between the two.

Modern societies proclaim the legal and moral equality of all men; so the difference in material conditions becomes less acceptable and raises a problem for the moralist and the statesman: to what extent are these differences legitimate? To what extent do they compromise the collective unity? In the past all societies have entailed hierarchies, 'classes' whose inequality was legally recognized. But because legal inequalities are officially eliminated and racial, national, and religious heterogeneities tend to disappear in certain nations, we ask ourselves about the inequalities still existing in the economic and social organization of modern societies. Hence the three questions issuing from reality itself: How do the different sorts of heterogeneity (income, occupation, standard of living, etc.) combine with one another? Are these groups more or less distinct according to social type? Are they pledged to co-operation or to struggle? Whatever the precision or imprecision of the possible answers, we cannot neglect these questions, because they are those which the societies of today in fact put to themselves.

We may invoke here Polanyi's distinction of logical levels:

'Natural science is regarded as a knowledge of things, while knowledge *about* science is held to be quite distinct from science, and is called "meta-science". We have then three logical levels: a first floor for the objects of science, a second for science itself and a third for meta-science, which includes the logic and epistemology of science. . . . A science dealing with living persons appears now logically different from a science dealing with inanimate things. In contrast to the two-storied logical structure of inanimate science, biological science, or at least some parts of biology, seem to possess a three-storied structure, similar to that of logic and epistemology. . . . Once we have before us the deliberate behaviour of an animal, by which it commits itself to a mode of action which can be right or wrong, and which thus implies assumptions about external things that can be true or false, the understanding of such a commitment is a theory of rightness and knowledge. It is clearly three-storied.'*

The study of living beings, as distinct from the study of inanimate objects, has a three-storied structure, because the animal may succeed or fail,

* *P.K.*, p. 344.

exhibit a true or false knowledge of reality, and so its behaviour evokes a theory of knowledge over and above the theory of practice (that is, of the living being in its spontaneous conduct).

By analogy, we may say that men in modern societies 'live' a certain distinction of classes. They live it implicitly in the manner in which they behave towards one another, and they live it consciously (and perhaps falsely in relation to their behaviour and their feelings) in their ideology of it. The sociologist does not give preference either to the level of practice or to that of ideology, but he seeks to grasp at one and the same time the level of experience and the ideological level, to put them both into a relation to the actual and historical facts, and finally to measure the gap between what this society wants to be and what it is (somewhat as the biologist ascertains the conformity of an individual to the norm of its species).

The understanding of a society by the sociologist, therefore, aims at a rigour, an exactitude, a coherence to which the spontaneous awareness that one of the members of the society has of it can make no claim, and yet such an understanding is by its nature nothing but the elaboration, through empirical concepts, of that same social consciousness. No discipline displays so clearly the precariousness of the opposition between objective and subjective meaning. The languages reconstructed by linguists were never thought of in this form by those who spoke the languages in question, and yet every linguistic system was present, included, in living speech. The sciences of man are essentially the elaboration, empirical and conceptual at once, of what is given in the universe experienced by men. There is no one single conceptual system implied by the existence of a society or a civilization, nor are there an unlimited number; still less is there a total freedom of choice.

(v) The verification of a proposition in physics by the facts, the 'validation' of an intellectual system, are always imperfect, subject to doubt and revision. Nor does any of the realms of culture exist other than in and through the community of those who elaborate it and live in it. While critical or empiricist thought erected a difference of kind between positive, demonstrated knowledge and moral, aesthetic or religious judgements, Michael Polanyi is concerned to re-establish the continuity between them by discovering once more the presence of the person, the fiduciary commitment even in the first steps in the acquisition of knowledge, in *learning*. Science, for him as for Bergson or Teilhard de Chardin, is a feature of the emergence of man and of mind in a cosmic and living process. At the same time, the critical task confronting man is to mark out the differences between the various realms of the spirit, once their fundamental kinship through reference to the person has been clarified. Thus the knowledge of the intellectual systems in which men have lived is itself elaborated into a system of knowledge. The history or the sociology of *Weltanschauungen* is personal knowledge, as are also the *Weltanschauungen* themselves, but the personal character is not the same in the two cases, and this is the difference which a theory of social knowledge must grasp and disentangle.

Max Weber had formulated this difference in a simple and categorical fashion. The knowledge which we gain of metaphysical systems is scientific because it points to a reality and because it takes account of facts and of observable and demonstrable relations. The systems of metaphysics themselves, on the other hand, lean towards a kind of knowledge which by its nature and scope exceeds the limits of what is accessible to science. Michael Polanyi cannot formulate the opposition in the same terms, because he has rediscovered trans-empirical beliefs within science itself, and because he perceives in the world a hierarchy of real orders and in the mind a plurality of realms of culture, none of these real orders being reducible to that which precedes it, and no authentic realm of culture being explicable by its external circumstances alone. Both the realm of conceptions of the world and the realm of the *knowledge* of conceptions of the world are authentic. In what respect and why can the last lay claim to a validity which the first does not achieve?

In one sense, the answer is simple: the conception of the world bears on realities external to man, though with ambitions and in a manner which do not accord with the demands of positive science. The *knowledge* of these conceptions of the world, on the other hand, bears on the realms of human life, on the realms of belief and faith in which the men of the past have lived. But this knowledge of human worlds must start from the meaning given to them by the men who have made their homes in them ('indwelling') and it cannot but accept, deny, or correct this meaning. In other words, the awareness of other human worlds is an element in a dialogue comparable to that which scientists have with one another or scientists with artists or priests. The knowledge of the sciences of man would no longer be a dialogue if all communion between the historian and the historical object were cancelled. The world of the historian is expressed in the knowledge that he acquires of the human worlds of the past, not only in the question he sets himself, but in the meaning he gives it. Even when the historian is seeking, as he must do, for the meaning which the dead had given to *their* world, he is interpreting, at least implicitly, this meaning in his own world. It is the historian of Auguste Comte, formed by the disciplines of positivism, Lucien Lévy-Bruhl, who created the ideal type of the primitive mentality.

But this answer, we must admit, is but a preface. True, critical thought and the primitive are engaged in dialogue, as is the Marxist with the liberal, Einstein with Newton, Hegel with Kant, the positivist with the Christian. But there are as many dialogues as there are spiritual worlds, and in addition to the dialogues within each world there are the dialogues of one world with another. To return to the case which interests us, that of the social sciences, the dialogue of the *Marxist* and the *liberal* is neither that of the positivist and the believer, nor that of Newton and Einstein.

A theory of the social sciences would develop from questions like these: how is economic or economic-sociological theory constructed? What is the role of elaboration comparable to that of mathematics? What is the role of political options or normative implications? What is the role, not of verification, but of experimental confirmation?

As an example, let me recall the refutation of the Marxist formulation by Schumpeter. If, he wrote, the rate of surplus value was such as Marx suggests, anyone at all employing human labour would derive benefits from so doing. A considerable gap between the value of wages, and the value created by work would open grandiose perspectives to all employers. On the other hand, the intervention of the average rate of profit, to conceal the gap between the surplus values of one branch or enterprise and the profits of that branch or that enterprise, is nothing more, in the eyes of Schumpeter, than the proof of the initial error in the theory. If the redistribution of profits does not answer to the expectations of the theory of surplus value, that is because the theory is not true. The average rate of profit is the equivalent of the so-called method of epicycles: supplementary hypotheses are multiplied in order to maintain the theory and to bring it into agreement with experience instead of starting over on a new foundation.

This argument, which I have very roughly summarized, fails to convince the majority of Marxists, even those who are not constrained by the police to unreason and orthodoxy. Probably agreement is wanting on the question, what is expected *of* or meant *by* theory. Probably the purely scientific intention, that of truth and universal validity, is crossed by political or metaphysical intentions. And yet it would be wrong to conclude from this that there is no truth in economic or sociological matters, and to affirm either that every scientist is the prisoner of his environment or that social consciousness is always the expression of historically particularized realities. The very person who formulates this proposition of thoroughgoing relativism is excepting himself in the very act of formulating it. What *is* true is that the sociologist's intention of stating universally valid truth is seldom entirely purified of all influence from historical subjectivity, and that even if such a purification were achieved the intention of the sociologist would still be imbued with historicity, because the framework of interpretation into which he sets his object inevitably stems from his own environment. But to infer from this a pure relativism, we should have to abstract from the *element* of fact which makes hypotheses or theories probable, improbable or absurd, as for instance, the facts make absurd every *schema* of capitalism which implies pauperization; and we should also have to abstract from the *coherence* appropriate to abstract theories like those of economics. Further, we should have to ignore the presence of the same facts expressed in a different language within different theories, and finally the possible ordering of the theories themselves at the hand of the social problems which the sociologist discovers even in his own society and in history.

If the sociologist were *not* radically distinguishable from the man in society, there would be no scientific dialogue, but only the contradictory justifications of the parties in the case. But if, on the other hand, the sociologist were in fact totally separate from the man in society, the scientific dialogue would not be charged with political or moral implications. The theorist has only to observe the actual development of the scientific dialogue to hear the two extreme interpretations. Social science is a factor in social man's awareness of his society (or societies). It arises with the intention of achieving universally valid and demonstrated truth, and proceeds to exceed the limits within which such an intention could be realized. The science of society not only gives to human collectives the consciousness they have of themselves and their destiny, but it clarifies each of them in relation to itself and to other such collectives. So long as the state does not stifle the aim of science by transforming scientists into mouthpieces of official truth, social science will trouble easy consciences by obliging every society to see what disfigures it and forbidding it to ignore what it is among other societies.

3

What, finally, is the relation between the leading ideas of *Personal Knowledge* and the epistemology of Max Weber? Let us enumerate Weber's fundamental propositions and notice briefly for each of them the answer that, as it seems to me, Polanyi has given to them, whether implicitly or explicitly.

(i) The sciences of culture are directed towards a specific kind of understanding, different from explanation as practised in physics and chemistry, since the meaning grasped by the interpreter is in a sense internal to reality. Polanyi admits that understanding is the proper aim of the sciences of culture, but he uses the same word, 'understand', for the intelligibility achieved by the sciences of nature. Without denying the peculiar traits of historical understanding, he sees in the latter a species of a genus, the ultimate term of an ascending series, rather than a break with the developments of the earlier sciences.

(ii) In the human world, unique events, singular facts are no less interesting, in fact they are more often of interest than are general propositions. Polanyi would subscribe to this statement. When the object is truly rare, as in the case of a great man, the understanding of that unique object can give the same satisfaction as a theory gives in physics when, in its simplicity and beauty, it accounts for many different phenomena.

(iii) The historian or the sociologist is more profoundly committed in his interpretation of the past and the present than is the physicist or the chemist in his explanation of motion or chemical composition. Polanyi agrees to this, but not without adding that trans-empirical, trans-rational commitment is present also even in the most basic procedures of positive science.

(iv) Despite this commitment, symbolized by the choice of questions or centres of interest, the historian or the sociologist wishes to achieve a rigorously objective truth, partial but universally valid. It is on this point that the fundamental opposition arises. Polanyi denies the antithesis between the arbitrary choice of values or interpretive systems and the scientific establishment of facts or relations. On the contrary, understanding is in its very nature the grasp of facts in a context; if the contexts are arbitrarily composed and only the facts universally valid, science will be arbitrary, like the contexts, and not universally valid like the facts.

(v) Finally, for Weber, facts and values would be rigorously heterogeneous and every hierarchy of values indemonstrable. Values would be contradictory and the battle of the gods would pledge humanity to tear itself apart. Here again, Polanyi answers that the understanding of works or persons involves appraisal. History or sociology does not cease to be scientific for including praise or blame. Understanding in biology is impossible without reference to the norms (trueness to type) of the living organism. Similarly, the sociologist's understanding of the statesman or the judge must refer to the morality of his act or the equity of his verdict.

But this comparison of epistemological principles will remain superficial unless we recall, in conclusion, the philosophical attitudes of Weber and Polanyi which we mentioned at the beginning of this essay.

That there was no science of values to be created or of conduct to be followed, Max Weber saw less as a check to reason than as a safeguard for the person. In marking off the limits of historical determinism, he was refuting the pretension of prophets to dictate our actions to us in the name of a future written in advance. In forbidding professors to introduce politics into their lecture rooms, he was preventing the confusion of demonstrations with preferences, and reserving the right to act according to conviction whatever the consequences. But in affirming an opposition of kind between procedures that are bound together in existence, between the discovery of the real and the choice of an action, he conferred an appearance of irrationality on decisions which indeed are not scientific, but which ought nevertheless to be reasonable. Having no intermediate category between science and choice, he finished by presenting, in terms of sheer choice, what tradition has more appropriately called wisdom.

And, as he denied all possibility of establishing rationally a hierarchy between the values of life and of the spirit, as the conflict of the gods of Olympus seemed to him symbolic of the incompatible values offered to or created by humanity, the choice of the man of action reflected, on a lower level, the higher choice of man, choice by each one of his destiny in the irreducible solitude of consciousness. But for such an unjustifiable choice, there is nothing to prevent its being the choice of the devil.

The commitment in which Max Weber's philosophy issues is a substitute for religious belief, the ultimate defence against a science which would violate our consciences by revealing their secrets or dictating their duties. But this last word of Max Weber is to be explained only by the hypothesis from which he had started. He accepted the objectivist vision of the world, ostensibly inspired by mechanistic science, before revolting against it. It is impossible to live in a world devoid of sense; if science eliminates religion and disenchants the world, then political and metaphysical commitment becomes, in its very irrationality, the expression of the condition set to man by the knowledge which explains the facts but strips the totality of meaning. But in such commitment, man escapes necessity, only to be lost in arbitrariness.

Michael Polanyi ignores or transcends these painful contradictions. The commitment of faith, he believes, is present from the first stage of knowledge and the hierarchy of spiritual worlds is ordered towards an ultimate goal. The existentialism of Polanyi issues in reconciliation because it moves towards religion. Thus the confrontation between Max Weber and Michael Polanyi over and above the epistemological problem, illustrates the dialogue of two persons: *the man of tragedy* and *the man of reconciliation*. The science of Polanyi leads without a break to faith; the science of Max Weber keeps a space for the faith which condemns it and which it denies.

11

Centre and Periphery

EDWARD SHILS

1

SOCIETY has a centre. There is a central zone in the structure of society. This central zone impinges in various ways on those who live within the ecological domain in which the society exists. Membership in the society, in more than the ecological sense of being located in a bounded territory and of adapting to an environment affected or made up by other persons located in the same territory, is constituted by relationship to this central zone.

The central zone is not, *as such*, a spatially located phenomenon. It almost always has a more or less definite location within the bounded territory in which the society lives. Its centrality has, however, nothing to do with geometry and little with geography.

The centre, or the central zone, is a phenomenon of the realm of values and beliefs. It is the centre of the order of symbols, of values and beliefs, which govern the society. It is the centre because it is the ultimate and irreducible; and it is felt to be such by many who cannot give explicit articulation to its irreducibility. The central zone partakes of the nature of the sacred. In this sense, every society has an 'official' religion, even when that society or its exponents and interpreters, conceive of it, more or less correctly, as a secular, pluralistic, and tolerant society. The principle of the Counter-Reformation: *Cuius regio, ejus religio*, although its rigor has been loosened and its harshness mollified, retains a core of permanent truth.

The centre is also a phenomenon of the realm of action. It is a structure of activities, of roles and persons, within the network of institutions. It is in these roles that the values and beliefs which are central are embodied and propounded.

2

The larger society appears, on a cursory inspection and by the methods of inquiry in current use, to consist of a number of interdependent sub-systems—the economy, the status system, the polity, the kinship system, and the institutions which have in their special custody the cultivation of cultural values, e.g. the university system, the ecclesiastical system,* etc. Each of these sub-systems itself comprises a network of organizations which are connected, with varying degrees of affirmation, through a common authority, overlapping personnel, personal relationships, contracts, perceived identities of interest, a sense of affinity within a transcendent whole, and a territorial location possessing symbolic value. (These sub-systems and their constituent bodies, are not equally affirmative *vis-a-vis* each other. Moreover the degree of affirmation varies through time, and is quite compatible with a certain measure of alienation within each *élite* and among the *élites*.)

Each of these organizations has an authority, an *élite*, which might be either a single individual or a group of individuals, loosely or closely organized. Each of these *élites* makes decisions, sometimes in consultation with other *élites* and sometimes, largely on its own initiative, with the intention of maintaining the organization, controlling the conduct of its members and fulfilling its goals. (These decisions are by no means always successful in the achievement of these ends, and the goals are seldom equally or fully shared by the *élite* and those whose actions are ordained by its decisions.)

The decisions made by the *élites* contain as major elements certain general standards of judgement and action, and certain concrete values, of which the system as a whole, the society, is one of the most pre-eminent. The values which are inherent in these standards, and which are espoused and more or less observed by those in authority, we shall call the *central value system* of the society. This central value system is the central zone of the society. It is central because of its intimate connection with what the society holds to be sacred; it is central because it is espoused by the ruling authorities of the society. These two kinds of centrality are vitally related. Each defines and supports the other.

The central value system is not the whole of the order of values and beliefs espoused and observed in the society. The value systems obtaining in any diversified society may be regarded as being distributed along a range. There are variants of the central value system running from hyper-affirmation of some of the components of the major, central value system to an extreme denial of some of these major elements in the central value system; the latter tends to but is not inevitably associated with, an affirmation of certain elements denied or subordinated in the central value system. There are also elements of the order of values and beliefs which are as random with respect to the central value system as the value and beliefs of human beings can be.†

* I use this term to include the religious institutions of societies which do not have a church in the Western sense of the term.

† There is always a considerable amount of unintegratedness of values and beliefs, both within the realm of value of representative individuals, and among individuals and sections of a society.

Centre and Periphery

The central value system is constituted by the values which are pursued and affirmed by the *élites* of the constituent sub-systems and of the organizations which are comprised in the sub-systems. By their very possession of authority, they attribute to themselves an essential affinity with the sacred elements of their society, of which they regard themselves as the custodians. By the same token, many members of their society attribute to them that same kind of affinity. The *élites* of the economy affirm and observe certain values which should govern economic activity. The *élites* of the polity affirm and observe certain values which should govern political activity. The *élites* of the university system and the ecclesiastical system affirm and practice certain values which should govern intellectual and religious activities (including beliefs). On the whole, these values are the values embedded in current activity. The ideals which they affirm do not far transcend the reality which is ruled by those who espouse them.* The values of the different *élites* are clustered into an approximately consensual pattern.†

One of the major elements in any central value system is an affirmative attitude towards established authority. This is present in the central value systems of all societies, however much these might differ from each other in their appreciation of authority. There is something like a 'floor', a minimum of appreciation of authority in every society, however liberal that society might be. Even the most libertarian and equalitarian societies that have ever existed possess at least this minimum appreciation of authority. Authority enjoys appreciation because it arouses sentiments of sacredness. Sacredness by its nature is authoritative. Those persons, offices or symbols endowed with it, however indirectly and remotely, are therewith endowed with some measure of authoritativeness.

The appreciation of authority entails the appreciation of the institutions through which authority works and the rules which it enunciates. The central value system in all societies asserts and recommends the appreciation of these authoritative institutions.

Implicitly, the central value system rotates on a centre more fundamental even than its espousal by and embodiment in authority. Authority is the agent of *order*, an *order* which may be largely embodied in authority

---

* This set of values corresponds to what the late Karl Mannheim called 'ideologies', i.e. values and beliefs, which are congruent with or embodied in current reality ('*seinskongruent*'). I do not wish to use the term 'ideology' to describe these value orientations. One of the most important reasons is that in the past few decades the term 'ideology' has been used to refer to intensely espoused value-orientations which are extremely *seinstranszendent*, which transcend current reality by a wide margin, which are explicit, articulated and hostile to the existing order. (For example, Bolshevist doctrine, National Socialist doctrine, Fascist doctrine, etc.) Mannheim called these 'utopias'. Mannheim's distinction was fundamental and I accept it, our divergent nomenclature notwithstanding.

† The degree of consensuality differs among societies and times. There are societies in which the predominant *élite* demands a complete consensus with its own more specific values and beliefs. Such is the case in modern totalitarian societies. Absolutist régimes in past epochs, which were rather indifferent about whether the mass of the population was party to a consensus, were quite insistent on consensus among the *élites* of their society.

or which might transcend authority and regulate it, or at least provide a standard by which existing authority itself is judged and even claims to judge itself. This order, which is implicit in the central value system, and in the light of which the central value system legitimates itself, is endowed with dynamic potentialities. It contains, above all, the potentiality of critical judgement on the central value system and the central institutional system.* The dynamic potentiality derives from the inevitable tendency of every concrete society to fall short of the order which is implicit in its central value system.

Closely connected with the appreciation of authority and the institutions in which it is exercised, is an appreciation of the *qualities* which qualify persons for the exercise of authority or which are characteristic of those who exercise authority. These qualities, which we shall call secondary values, can be ethnic, educational, familial, economic, professional; they may be ascribed to individuals by virtue of their relationships or they may be acquired through study and experience. But whatever they are, they enjoy the appreciation of the central value system simply because of their connection with the exercise of authority. (Despite their ultimately derivative nature, each of them is capable of possessing an autonomous status in the central zone, in the realm of the sacred; consequently, severe conflicts can be engendered.)

The central value system thus comprises secondary as well as primary values. It legitimates the existing distribution of roles and rewards to persons possessing the appropriate qualities which in various ways symbolize degrees of proximity to authority. It legitimates these distributions by praising the properties of those who occupy authoritative roles in the society, by stressing the legitimacy of their incumbency of those roles, and the appropriateness of the rewards they receive. By implication, and explicitly as well, it legitimates the smaller rewards received by those who live at various distances from the circles in which authority is exercised.

The central institutional system may thus be described as the set of institutions which is legitimated by the central value system. Less circularly, however, it may be described as those institutions which, through the radiation of their authority, give some form to the life of a considerable section of the population of the society. The economic, political, ecclesiastical and cultural institutions impinge compellingly at many points on the conduct of much of the population in any society through the actual exercise of authority and the potential exercise of coercion, through the provision of persuasive models of action, and through a partial control of the allocation of rewards. The kinship and family systems, although they have much smaller radii, are microcosms of the central institutional system, and do much to buttress its efficacy.

---

* To use Mannheim's terminology, while going beyond Mannheim, every 'ideology' has within it a 'utopian' potentiality. To use my own terminology, every central value system contains within itself an ideological potentiality.

3

The existence of a central value system rests, in a fundamental way, on the need which human beings have for incorporation into something which transcends and transfigures their concrete individual existence. They have a need to be in contact with symbols of an order which is larger in its dimensions than their own bodies and more central in the 'ultimate' structure of reality than is their routine everyday life. Just as friendship exists because human beings must transcend their own self-limiting individuality in personal communion with another personality, so membership in a political society is a necessity of man's nature.* There is need to belong to a polity just as there is a need for conviviality. Just as a person shrivels, contracts, and corrupts when separated from all other persons or from those persons who have entered into a formed and vital communion with him, so the man with political needs is crippled and numbed by his isolation from a polity or by his membership in a political order which cannot claim his loyalty.

The need for personal communion is a common quality among human beings who have reached a certain level of individuation. Those who lack the need and the capacity impress us by their incompleteness. The political need is not so widely spread or highly developed in the mass of the population of any society as the need and capacity for conviviality. Those who lack it impress by their 'idiocy'. Those who possess it add the possibility of civility to the capacity for conviviality which we think a fully developed human being must possess.

The political need is of course nurtured by tradition but it cannot be accounted for by the adduction of tradition. The political need is a capacity like certain kinds of imagination, reasoning, perceptiveness, or sensitivity. It is neither instinctual nor learned. It is not simply the product of the displacement of personal affects on to public objects, although much political activity is impelled by such displacement. It is not learned by teaching or traditional transmission, though much political activity is guided by the reception of tradition. The pursuit of a political career and the performance of civil obligations gains much from the impulsion of tradition. None the less, tradition is not the seed of this inclination to attach oneself to a political order.

The political need, which may be designated as the need for civility, entails sensitivity to an order of being where 'creative power' has its seat. This creative centre which attracts the minds of those who are sensitive to it is manifested in authority operating over territory. Both authority and territory convey the idea of potency, of 'authorship', of the capacity to do

---

* This by no means implies that this satisfaction of the need to be a member of a transcendent body, be it a tribe or a nation or a political community, exhausts the functions of political community. A political community performs many functions and satisfies many needs which have little to do with the need for membership in a political community.

vital things, of a connection with events which are intrinsically important. Authority is thought, by those with the political or civil need, to possess this vital relationship to the centre from which a right order emanates. Those who are closely and positively connected with authority, through its exercise or through personal ties, are thought, in consequence of this connection, to possess a vital relationship to the centre, the locus of the sacred, the order which confers legitimacy. Land—'territoritiality'—has similar properties, and those who exercise authority through control of land have always been felt to enjoy a special status in relation to the core of the central value system. Those who live within given territorial boundaries come to share in these properties and thus become the objects of political sentiments. Residence within certain territorial boundaries, and rule by common authority are the properties which define citizenship and establish its obligations and claims.*

It must be stressed that the political need is not by any means equally distributed in any society, even the most democratic. There are human beings whose sensitivity to the ultimate is meagre, although there is perhaps no human being from whom it is entirely absent. Nor does sensitivity to remote events which are expressive of the centre always focus on their manifestations in the polity.

Apolitical scientists who seek the laws of nature but are indifferent, except on grounds of prudence, to the laws of society are one instance of this uneven development of sensitivity to ultimate things. Religious persons who are attached to transcendent symbols without embodiment in civil polity or in ecclesiastical organization represent another variant. In addition to these, there are very many persons whose sensitivity is exhausted long before it reaches so far into the core of the central value system. Some have a need for such contact only in crises and on special periodic occasions, at the moment of birth or marriage or death, or on holidays. Like the intermittent, occasional, and unintense religious sensibility, the political sensibility, too, can be intermittent and unintense. It might come into operation only on particular occasions, e.g. at election time, or in periods of severe economic deprivation or during a war or after a military defeat. Beyond this there are some persons who are never stirred, who have practically no sensibility as far as events of the political order are concerned.

Finally, there are persons, not many in any society but often of great importance, who have a very intense and active connection with the centre, with the symbols of the central value system, but whose connection is passionately negative.* Equally important are those who have a positive but

---

* It is not entirely an accident that nationalism is connected with land reform. Land reform is part of a policy which seeks to disperse the special relationship to a higher order of being from a few persons, that is, the great landlords in whom it was previously thought to be concentrated, to the large mass of those who live upon the territory.

* T. S. Eliot has pointed out, in discussing Baudelaire, the profound difference between the atheist who feels strongly about the nature of the universe and who is vehemently antireligious and the person who is utterly indifferent to religion.

no less intense and active connection with the symbols of the centre, a connection so acute, so pure, and vital that it cannot tolerate any falling short in daily observance such as characterizes the *élites* of the central institutional system. These are often the persons around whom a sharp opposition to the central value system and even more to the central institutional system is organized. From the ranks of these come prophets, revolutionaries, doctrinaire ideologists for whom nothing less than perfection is tolerable.

4

The need for established and created order, the respect for creativity and the need to be connected with the 'centre' do not exhaust the forces which engender central value systems. To fill out the list, we must consider the nature of authority itself. Authority has an expansive tendency. It has a tendency to expand the order which it represents towards the saturation of territorial space. The acceptance of the validity of that order entails a tendency towards its universalization within the society over which authority rules. Ruling indeed consists in the universalization within the boundaries of society, of the rules inherent in the order. Rulers, simply out of their possession of authority and the impulses which it generates, wish to be obeyed and they wish to obtain assent to the order which they symbolically embody. The symbolization of order in offices of authority has a compelling effect on those to whom authority is directed; it has an even more compelling effect on those who occupy those offices.

In consequence of this, rulers seek to establish a universal diffusion of the acceptance and observance of the values and beliefs of which they are the custodians through incumbency in those offices. They use their powers to punish those who deviate and to reward with their favour those who conform. Thus, the mere existence of authority in society imposes a central value system on that society.*

Not all persons who come into positions of authority possess the same responsiveness to the inherently dynamic and expansive tendency in authority. Some are more attuned to it; others are more capable of resisting it. Tradition, furthermore, acts as a powerful brake upon expansiveness, as does the degree of differentiation of the structure of *élites* and of the society as a whole.

5

The central institutional system, probably even in revolutionary crises, is the object of a substantial amount of consensus. The central value system

---

* I would regret an easy misunderstanding to which the sentences above might give rise. There is much empirical truth in the common observations that rulers 'look after their own', that they are only interested in remaining in authority, in reinforcing their possession of authority and in enhancing their security of tenure through the establishment of a consensus built around their own values and beliefs. None the less these observations seem to me to be too superficial. They fail to discern the dynamic property of authority as such, and particularly of authority over society.

which legitimates the central institutional system is widely shared, but the consensus is never perfect. There are differences within even the most consensual society about the appreciability of authority, the institutions within which it resides, the *élites* which exercise it and the justice of its allocation of rewards.

Even those who share in the consensus, do so with different degrees of intensity, whole-heartedness and devotion. As we move from the centre of society, the centre in which authority is possessed, to the hinterland or the periphery, over which authority is exercised, attachment to the central value system becomes attenuated. The central institutional system is neither unitary nor homogeneous, and some levels have more majesty than others. The lower one goes in the hierarchy, or the further one moves territorially from the locus of authority, the less likely is the authority to be appreciated. Likewise, the further one moves from those possessing the secondary traits associated with the exercise of authority into sectors of the population which do not equally possess those qualities, the less affirmative is the attitude towards the reigning authority, and the less intense is that affirmation which does exist.

Active rejection of the central value system is, of course, not the sole alternative to its affirmation. Much more widespread, in the course of history and in any particular society, is an intermittent, partial and attenuated affirmation in the central value system.

For the most part, the mass of the population in pre-modern societies have been far removed from the immediate impact of the central value system. They have possessed their own value systems which were *occasionally* and *fragmentarily* articulated with the central value system. These pockets of approximate independence have not, however, been completely incompatible with isolated occasions of articulation and of intermittent affirmation. Nor have these intermittent occasions of participation been incompatible with occasions of active rejection and antagonism to the central institutional system, to the *élite* which sits at its centre, and to the central value system which that *élite* puts forward for its own legitimation.

The more territorially dispersed the institutional system, the less the likelihood of an intense affirmation of the central value system. The more inegalitarian the society, the less the likelihood of an intense affirmation of the central value system, especially where, as in most steeply hierarchical societies, there is a large and discontinuous gap between those at the top and those below them. Indeed, it might be said that the degree of affirmation inevitably shades off from the centre of the exercise of authority and of the promulgation of values.

As long as societies were loosely co-ordinated, as long as authority lacked the means of intensive control, and as long as much of the economic life of the society was carried on outside any market or almost exclusively in local markets, the central value system invariably became attenuated in the outlying reaches. With the growth of the market, and the administrative and technological strengthening of authority, contact with the central value system increased.

When, as in modern society, a more unified economic system, political democracy, urbanization, and education have brought the different sections of the population into more frequent contact with each other and created even greater mutual awareness, the central value system has found a wider acceptance than in other periods of the history of society. At the same time these changes have also increased the extent, if not the intensity, of active 'dissensus' or rejection of the central value system.

The same objects which previously engaged the attention and aroused the sentiments of a very restricted minority of the population have in modern societies become concerns of much broader strata of the population. At the same time that increased contact with authority has led to a generally deferential attitude, it has also run up against the tenacity of prior attachments and a reluctance to accept strange gods. Class conflict in the most advanced modern societies is probably more open and more continuous than in pre-modern societies but it is also more domesticated and restricted by attachments to the central value system.*

The old gods have fallen, religious faith has become much more attenuated in the educated classes and suspicion of authority is much more overt than it has ever been. Nonetheless in the modern societies of the West, the central value system has gone much more deeply into the heart of their members than it has ever succeeded in doing in any earlier society. The 'masses' have responded to their contact with a striking measure of acceptance.

6

The power of the ruling class derives from its incumbency of certain key positions in the central institutional system. Societies vary in the extent to which the ruling class is unitary or relatively segmental. Even where the ruling class is relatively segmental, there is, because of centralized control of appointment to the most crucial of the key positions or because of personal ties or because of overlapping personnel, some sense of affinity which, more or less, unites the different sectors of the *élite*.†

---

* Violent revolutions and bloody civil wars are much less characteristic of modern societies than of pre-modern societies. Revolutionary parties are feeble in modern societies which have moved towards widespread popular education, a greater equality of status, etc. The strength of revolutionary parties in France and Italy are a measure of the extent to which French and Italian societies have not become modernized. The inertness, from a revolutionary point of view, of their rank and file is partially indicative of the extent to which, despite their revolutionary affiliations, the working class has become assimilated into the central value system of their respective societies.

† The segmentation or differentiation in the structure of *élites* is an important factor in limiting the expansiveness of authority among the *élites*. A differentiated structure of *élites* brings with it a division of powers, which can be radically overcome only by draconic measures. It can be done, as the Soviet Union has shown, but it is a perpetual source of strain, as recent Soviet developments have likewise shown.

This sense of affinity rests ultimately on the high degree of proximity to the centre which is shared by all these different sectors of the ruling class. They have, it is true, a common vested interest in their position. It is not, however, simply the product of a perception of a coalescent interest; it contains a substantial component of mutual regard arising from a feeling of a common relationship to the central value system.

The different sectors of the *élite*, even in a highly pluralistic society where the *élite* is relatively segmental in its structure, are never equal. One or two usually predominate, to varying degrees, over the others, even in situations where there is much mutual respect and a genuine sense of affinity. Regardless, however, of whether they are equal or unequal, unitary or segmental, there is usually a fairly large amount of consensus among the *élites* of the central institutional system. This consensus has its ultimate root in their common feeling for the transcendent order which they believe they embody or for which they think themselves responsible.*

7

The mass of the population in all large societies stand at some distance from authority. This is true both with respect to the distribution of authority and the distribution of the secondary qualities associated with the exercise of authority.

The functional and symbolic necessities of authority require some degree of concentration. Even the most genuinely democratic society, above a certain very small size, requires some concentration of authority for the performance of elaborate tasks. It goes without saying that non-democratic societies have a high concentration of authority. Furthermore, whether the society is democratic or oligarchical, the access to the key positions in the central institutional system tends to be confined to persons possessing a distinctive constellation of properties, such as age, educational, ethnic, regional, and class provenience, etc.

The section of the population which does not share in the exercise of authority and which is differentiated in secondary properties from the exercisers of authority, is usually more intermittent in its 'possession' by the central value system. For one thing, the distribution of sensitivity to remote, central symbols is unequal, and there is a greater concentration of such sensitivity in the *élites* of the central institutional system. Furthermore, where there is a more marginal participation in the central institutional system, attachment to the central value system is more attenuated. Where the central institutional system becomes more comprehensive and inclusive so that a larger proportion of the life of the population comes

---

* This does not obtain equally for all *élites*. Some are much more concerned in an almost entirely 'secular' or manipulative way with remaining in power. Nonetheless, even in a situation of great heterogeneity and much mutual antipathy, the different sectors of the *élite* tend to experience the 'transforming' transcendental overtones which are generated by incumbency in authoritative roles, or by proximity to 'fundamentally important things'.

within its scope, the tension between the centre and the periphery, as well as the consensus, tends to increase.

The mass of the population in most pre-modern and non-Western societies have in a sense lived *outside* society and have not felt their remoteness from the centre to be a perpetual injury to themselves. Their low position in the hierarchy of authority has been injurious to them, and the consequent alienation has been accentuated by their remoteness from the central value system. The alienation has not however been active or intense, because, for the most part, their convivial, spiritual and moral centre of gravity has lain closer to their own round of life. They have been far from fullfledged members of their societies and they have very seldom been citizens.

Among the most intensely sensitive or the more alertly intelligent, their distance from the centre accompanied by their greater concern with the centre, has led to an acute sense of being on 'the outside', to a painful feeling of being excluded from the vital zone which surrounds 'the centre' of society (which is the vehicle of 'the centre of the universe'). Alternatively these more sensitive and more intelligent persons have, as a result of their distinctiveness, often gained access to some layer of the centre by becoming school-teachers, priests, administrators. Thus they have entered into a more intimate and more affirmative relationship with the 'centre'. They have not in such instances, however, always overcome the grievance of exclusion from the most central zones of the central institutional and value systems. They have often continued to perceive themselves as 'outsiders', while continuing to be intensely attracted and influenced by the outlook and style of life of the centre.

8

Modern large-scale society rests on a technology which has raised the standard of living and which has integrated the population into a more unified economy. In correspondence with these changes, it has witnessed a more widespread participation in the central value system through education, and in the central institutional system through the franchise and mass communication. On this account, it is in a different position from all premodern societies.

In modern society, in consequence of its far greater involvement with the central institutional system, especially with the economy and the polity, the mass of the population is no longer largely without contact with the central value system. It has, to an unprecedented extent, come to feel the central value system to be its own value system. Its generally heightened sensitivity has responded to the greater visibility and accessibility of the central value system by partial incorporation. Indeed, although, compared with the *élite*, its contact is still relatively intermittent and unintense, that enhanced frequency and intensity are great universal-historical novelties. They are nothing less than the incorporation of the mass of the population into society. The 'process of civilization' has become a reality in the modern world.*

To a greater extent than ever before in history the mass of the population in modern Western societies feel themselves to be 'part' of their society in a way in which their ancestors never did. Just as they have become 'alive' and hedonistic, more demanding of respect and pleasure, so, too, they have become more 'civilized'. They have come to be parts of the civil society with a feeling of attachment to that society and a feeling of moral responsibility for observing its rules and for sharing in its authority. They have ceased to be primarily objects of authoritative decisions by others; they have become to a much greater extent, acting and feeling subjects with wills of their own which they assert with self-confidence. Political apathy, frivolity, vulgarity, irrationality, and responsiveness to political demagogy are all concomitants of this phenomenon. Men have become citizens in larger proportions than ever before in the large states of history, and probably more, too, than in the Greek city states at the height of the glory of their aristocratic democracies.

The emergence of nationalism, not just the fanatical nationalism of politicians, intellectuals, and zealots, but as a sense of nationality and an affirmative feeling for one's own country, is a very important aspect of this process of the incorporation of the mass of the population into the central institutional and value systems. The more passionate type of nationalism is an unpleasant and heroic manifestation of this deeper growth of civility.

9

None the less this greater incorporation carries with it also an inherent tension. Those who participate in the central institutional and value systems—who feel sufficiently closer to the centre now than their forebears ever did—also feel their position as outsiders, their remoteness from the centre, in a way in which their forebears probably did not feel it.†

* Cf. Elias, Norbert, *Der Prozess der Zivilisation*, 2 vols. Basel, 1937. The phenomenon of '*das sinkende Kulturgut*' was noticed by German writers on late medieval society, and a parallel phenomenon was observed by Max Weber in his studies of Indian society. He called it 'Brahmanization'. This theme has been well treated by Professor M. N. Srinivas in his studies of 'Sanskritization'. This assimilation of elements of the value systems of higher classes and castes by lower strata which occurs in every society is not, however, identical either in quality or extent with the growth of the *sense of fundamental affinity* which characterizes modern society.

† The modern trade union movement, which has disappointed those whose revolutionary hopes were to be supported by the organized working classes, illustrates this development. The leadership of the trade unions have come to be part of the central institutional system and accordingly, at least in part they fulfil the obligations which are inherent in action within that system. At the same time, their rank and file members also have come to share more widely and intensely in the central value system and to affirm more deeply and continuously than in the past the central institutional system. Nonetheless, the leadership of the trade unions, deriving from sections of the society which have felt themselves to be 'outside' of the prevailing society, still and necessarily, carry traces of that position in their outlook; their rank and file, less involved in the central institutional system than the leadership, experience even more acutely their position as 'outsiders' *vis-à-vis* the central value system. The more sensitive among them are the most difficult for the leaders of the unions to hold in check.

Parallel with this incorporation of the mass of the population into society—halting, spotty, and imperfect as this incorporation is—has gone a change in the attitudes of the ruling classes of the modern states of the West. (In Asia and Africa, the process is even more fragmentary, corresponding to the greater fragmentariness of the incorporation of the masses into those societies.) In the modern Western states, the ruling classes have come increasingly to acknowledge the dispersion, into the wider reaches of the society, of the *charisma* which informs the 'centre'. The qualities which account for the expansiveness of authority have come to be shared more widely in the population, quite far from the 'centre' where reside the incumbents of the positions of authority. In the eyes of the *élites* of the modern states of the West, the mass of the population have somehow come to share in the vital connection with the 'order' which inheres in the central value system and which was once thought to be in the special custody of the ruling classes.

The *élites* are, of course, more responsive to sectors of society which have voting power, and therewith, legislative power, and which possess agitational and purchasing powers as well. These would make them simulate respect for the populace even where they did not feel it. Nonetheless, mixed with this simulated respect, is also a genuine respect for the mass of the population as bearers of a true individuality and a genuine, even if still limited appreciation of their intrinsic worth as fellow-members of the civil society and, in the deepest sense, vessels of the *charisma* which lives at the 'centre' of society.*

10

There is a limit to consensus. However comprehensive the spread of consensus, it can never be all embracing. A differentiated large-scale society will always be compelled by professional specialization, tradition, the normal distribution of human capacities and an inevitable antinomianism to submit to inequalities in participation in the central value system. Some persons will always be a bit closer to the centre, some will always be more distant from the centre.

None the less, the expansion of individuality attendant on the growth of individual freedom, and opportunity and the greater density of communications have contributed greatly to narrowing the range of inequality. The

---

* The populism of the rulers of totalitarian and oligarchical societies is, in part, hypocrisy, and, in part, acknowledgement of the existence of 'outsider' feelings in these *élites*, who still believe in their hearts that the modern liberal Western states constitute 'the centre' of the world. But I would venture to state that there is more to it than that. These oligarchical and totalitarian *élites* also share in this fundamental expansion of sensibility and empathy which opens their imaginations to the *charisma* of the ordinary human beings who live outside the key positions of the central institutional system. These observations should not, however, obscure the fact that this widened sensibility coexists with a still very deeply rooted belief in the concentration of *charisma* in the authoritative centre of society. Their widened sensibility must still contend with their appreciation of the sacredness of the peaks of authority in the central institutional system.

peak at the centre is no longer so high, the periphery is no longer so distant.

The individuality which has underlain the entry into the consensus around the central value system might in the end also be endangered by it. Liberty and privacy live on islands in a consensual sea. When the tide rises they may be engulfed. This is another instance of the dialectical relationships among consensus, indifference, and alienation, but further consideration must be left for another occasion.

BIBLIOGRAPHICAL NOTE

I have been attempting to develop the line of thought which is presented here in a number of studies to which an interested reader might refer:

1. 'The Meaning of the Coronation' (with Michael Young), *The Sociological Review*, December 1953.
2. 'Primordial, Personal, Sacred, and Civil Ties,' *The British Journal of Sociology*, June 1957.
3. 'Ideology and Civility,' *The Sewanee Review*, Summer 1958.
4. 'The Concentration and Dispersion of Charisma,' *World Politics*, October 1958.
5. 'Political Development in New States,' *Comparative Studies in History and Society*, Spring and Summer 1960.
6. 'Mass Society and its Culture, *Daedalus*, Spring, 1960.
7. 'Metropolis and Province in the Intellectual Community,' *Kenyon Review*, Winter 1961.

12

The Republic of Science

BERTRAND DE JOUVENEL

1

MICHAEL POLANYI has moved from a great career as a scientist to a great career as a philosopher. Given the exceptional powers required for this duality of achievements, the former experience afforded exceptional material for the latter preoccupation. The great question 'How do I think?' could be approached by him in the following guise: 'When my attention was addressed to understanding the external world, how did I proceed to form hard-earned judgements, and what was implied in their formation?' Thus Polanyi the philosopher was not observing Polanyi the philosopher but Polanyi the scientist.

In so doing Polanyi has shattered important prejudices about the procedure of scientific discovery, and drawn an unforgettable picture of the hunting mind tracking down its quarry. Thus described in the very act of its chase, the human mind best reveals the nature of its powers: the impression gained by this reader at least is that the creative mind is wilder and less articulate than we were wont to think. Be that as it may, Polanyi has naturally progressed from a renovated conception of intellectual activity to a renovated image of man.

Such renovation must exercise a deep, pervading influence upon political philosophy; this cannot yet have taken place, and I am not so bold as to forecast its form. I have chosen a far less ambitious theme. Leaving to others the main stream of Polanyi's thinking I shall devote my remarks to an arm branching out therefrom.

In the Riddell Memorial Lectures,[1] Polanyi calls attention to the structure and nature of the 'community of scientists', a subject to which he returns more briefly in the *Logic of Liberty*[2] and which is again touched upon in *Personal Knowledge*.[3] In the first and most extensive statement, it appears clearly that Polanyi regards what we may call the unwritten constitution of the Republic of Science as not only interesting in itself but also suggestive for political science. Recent conversation with Polanyi, while leaving him unaware that I proposed to study this small segment of his work, satisfied me that, now as well as in 1946, he does regard his analysis as providing a good starting point for a discussion of political organization. I shall therefore venture into this 'place' staked out by him, hoping that my clumsy spade-work may provoke him to the far more competent exploitation which we can expect from him.

Thus my purpose is well delimited: it is to concentrate upon a concise description of the Republic of Science[4] which, astonishingly enough is, as far as I know, the only one in existence, and which, not at all astonishingly, is brimming with important suggestions.

2

One enters civil society by mere birth, and one becomes a citizen by mere coming of age. Not so in the Republic of Science, wherein membership must be diligently sought and is selectively granted. Polanyi characterizes the stages through which the candidate must travel. At school, he must acquire 'a facility in using scientific terms to indicate the established doctrine, the dead letter of science'.[5] All that is then asked of him is that he should incorporate into his mind the 'scientific verities' which are to be found in textbooks, and acquire a certain agility in applying general statements to particular cases: this proves hard enough, many fall by the way. The university forms the second stage: the same process is further pursued but in a different spirit: 'The university tries to bring this knowledge to life by making the student realize its uncertainties and its eternally provisional nature, and giving him perhaps a glimpse of the dormant implications which may yet emerge from the established doctrine. It also imparts the beginnings of scientific judgement by teaching the practice of experimental proof and giving a first experience in routine research.'[6] The student emerges from the university with a degree, i.e. he has passed a 'means test'. Such achievement has required a prolonged application of will; sad cases, however, testify that will is not enough, some gifts are required. Thus the process of scientific education weeds out both the insufficiently gifted and the insufficiently willing.

But even at the end of this long and difficult process, the graduate has no voice or vote in the scientific community. He has no voice. When a leading politician makes a statement, any adult inhabitant can utter a counter-statement and, if he brings to it sufficient vigour, he can collect an audience to hear his views. It does not matter that the statement attacked relates to some complex problem, such as defence, or fiscal and monetary policy: it is not required of the protesting citizen that he should produce formal evidence of reliable knowledge acquired by him in the field under discussion. If now we shift to the Republic of Science, we find things very different indeed: even those who can show degrees testifying that they have acquired reliable knowledge cannot get a hearing if they wish to protest against a statement, however surprising, made by a leading scientist. They do not doubt the propriety of leaving such a statement to be debated among 'their betters'.

Nor are those to whom such discussion is left their elected representatives. It should be noted, firstly, that key positions in the world of science are not filled by the majority vote of graduates, secondly that 'authorities' in the field are not of necessity identical with position holders, thirdly that such authorities are recognized as such by virtue of their previous performances in research and discovery, performances which themselves have been warranted not by a majority verdict of the large body of graduates but by a process of acknowledgement which depends upon a far lesser number of people, and which is itself complex. The men who are deemed competent to debate a challenging statement can be called 'representative', provided one contrasts the meaning here implied to the meaning current in politics. These are not representative of the *populus* of graduates, made such by a majority vote, but they are representative of Science, to which this *populus* is dedicated, and they are acknowledged as such by reason of their works.

The qualification to intervene in a scientific debate is earned by positive achievement in the field. Polanyi indicates that the most common pathway to such positive achievement runs through apprenticeship to a senior:

'A full initiation into the premises of science can be gained only by the few who possess the gifts for becoming independent scientists, and they usually achieve it only through close personal association with the intimate views and practice of a distinguished master. In the great schools of research are fostered the most vital premises of scientific discovery. A master's daily labours will reveal these to the intelligent student and impart to him also some of the master's personal intuitions by which his work is guided. The way he chooses problems, selects a technique, reacts to new clues and unforseen difficulties, discusses other scientists' work and keeps speculating all the time about a hundred possibilities which are never to materialize, may transmit a reflection at least of his essential visions. This is why so often great scientists follow great masters as apprentices. Rutherford's work bore the clear imprint of his apprenticeship under J. J. Thompson. And no less than four Nobel laureates are found in turn among the personal pupils of Rutherford, Some forms of science, such as psychoanalysis, can hardly be transmitted by precept. Every psychoanalyst today has either been analysed by Freud or by another psychoanalyst who has been so analysed, etc. (Perhaps a modern version of the Apostolic Succession.) Research in the chemistry of carbohydrates in Britain has been almost entirely the work of four scientists, Purdy, Irvine, Haworth, and Hirst, who followed each other in single file as master and pupils.'[6]

3

Polanyi thus shows that the Republic of Science has a highly aristocratic structure. But further, membership in the community of scientists requires commitment to a common faith. No one can become a scientist who is not driven by a primary urge for discovery, who is not the ardent suitor of a hidden beauty. Somewhat romantically, scientists can be likened to a company of knights dispersed in search of Sleeping Princesses, all of whom are more or less distantly related. The spirit of the quest is essential to the making of a scientist, and forms a fundamental bond between scientists. But however necessary, such impulsive love is not sufficient. Another image comes to mind, so suitable that I shall use it even though it combines badly with the first.

Cervantes, if I rightly remember, describes Quixote fixing up a helmet for his adventuring. He spends a good deal of trouble in putting it together: when he has done, he deals it a great blow to see whether it will stand. The helmet is shattered. Quixote then begins all over again, and when the product is finished, he reaches for his sword, to test the value of his work. But on second thoughts, he fears a second disaster, and dons the helmet without further trial. Now his attitude in the second instance, *mutatis mutandis*, would disqualify a scientist.

Polanyi stresses the fact that a great deal of guessing is involved in scientific discovery. The very inception of a research project rests upon the guess that an unknown pattern links together different phenomena, and the first steps are further guided by guesses narrowing down the form of the pattern to a few alternatives; long before the successive steps have all been taken, the scientist will have come to believe that the pattern has a certain definite shape. This, however, is not a finding: it can only be taken as an assumption from which the searcher will have to work back, making sure that he has a properly consecutive series of steps, and that the facts fit into the pattern. If he fails to pursue this testing with great rigour, he may still possibly have hit upon a truth but he will have fallen from the standards of scientific inquiry. To a scientist, such standards are not rules imposed upon him from the outside by the society to which he belongs, but they have become in-built and cherished parts of his personality. A scientist does not chafe under the obligation of verifying and tightening up his demonstration, he enjoys it, and in many cases lingers over this enjoyment in such a manner as to be overtaken by some colleague who publishes sooner. But any scientist would feel ashamed to break his unspoken vows. The grace which has touched him and impelled him towards research binds him to the works which are seemly. It is because such dedication is common to all scientists that they form a community, and not the other way round:

'The devotion of all scientists to the ideals of scientific work may be regarded as the General Will governing the society of scientists. But this identification makes the General Will appear in a new light. It is seen to differ from any other will by the fact that it cannot vary its own purpose. Scientists who would suddenly lose all passion for science, and take up instead an interest in greyhounds, would instantly cease to form a scientific society. The co-operative structure of scientific life could not serve the purpose of the joint breeding of greyhounds, for the pursuit of which the former scientists would have to organize themselves once more quite afresh. Scientific society is not and cannot be formed by a group of persons taking first the decision of binding themselves to a General Will and then choosing to direct their general will to the advancement of science. Scientific life illustrates, on the contrary, how the general acceptance of a *definite* set of principles brings forth a community governed by these principles—a community which would automatically dissolve the moment its constitutive principles were repudiated. The General Will appears then as a rather misleading fiction; the truth being (if the case of science be a guide) that voluntary submission to certain principles necessarily generates a communal life governed by these principles, and that ultimate sovereignty then rests safely with each generation of individuals who, in their devotion to these principles, conscientiously interpret and apply them to the issues of the period.

'This also throws new light on the nature of the Social Contract. In the case of the scientific community the contract consists in the gift of one's own person—not to a sovereign ruler as Hobbes thought, nor to an abstract General Will as Rousseau postulated—but to the service of a particular ideal.'[7]

4

The foregoing statement calls for extensive comment. Not that it requires any elucidation, it could not be clearer, but because comment is an occasion of dwelling upon an author's expressions, and achieving deeper contact with his thought. Whether Polanyi does somewhat less than justice to either Hobbes or Rousseau is here immaterial; we are concerned here not with what they meant but with what he means. His reason for referring to them is plain. All Social Contract writers stress the intention of individuals to coalesce into a body politic. If I may use an ugly word, 'togetherness' is sought in itself. Such is not the case of the scientific body. Here individuals severally dedicate themselves to the service of Science and, as it were, find themselves in association because of their identical devotions. The implications of the contrast are many and far-reaching.

As there is no doubt that the social state is indispensable to the satisfaction of every human need,* Social Contract authors have found no difficulty in naming a variety of reasons which can motivate men to form a Society (as they put it) or to maintain its existence. The most plausible reason for regarding it arbitrarily as originating in a meeting of wills is to emphasize the threat to it involved in the divorce of wills.

Wilful as men are, they can be expected to use the opportunities afforded by the social state, and indeed the collective powers if available to them, for the furtherance of their several ends, inspired by their interests and passions. Hobbes and Rousseau were both extremely fearful of the resulting strife: the former saw but one remedy for it, an absolute power placed beyond the reach of factions, the latter saw 'but one alternative to Hobbesism', as he put it, a disposition of individuals such that all should wish to feel as one, thus setting a supreme value upon community itself. Rousseau added the discouraging rider that such moral unity cannot be preserved in

---

* This had led the previous authors to regard Society as the given milieu within which individuals take shape, to me a more realistic approach than the social contract theory.

a large and advanced society. Polanyi suggests the example of the scientific community, wherein a satisfactory degree of harmony is assured, not by a central power, not because members are dedicated to the community, but because they are individually dedicated to something which stands far above this community, and for the sake of which this community functions.

It is clear enough that he regards this model as valid beyond the case of the scientific community: this comes out in a sentence of the last quotation: 'the truth being (if the case of science be a guide) that voluntary submission to certain principles necessarily generates a communal life governed by these principles.'

Here it is indicated that the case of science is offered as a guide to a general proposition: 'voluntary submission to certain principles necessarily generates a communal life . . .' and it is further suggested that this proposition is susceptible of application beyond the realm of the scientific community. That the proposition is true and important seems to me beyond doubt. It has considerable analytical power, leading us to distinguish and contrast bodies which are based upon a common dedication to some principle transcending the body, from those bodies which are not so based. This contrast we shall now elaborate, using the opposing terms of civil society and dedicated company. This will help us to discuss subsequently whether the model of the dedicated company can with proper adjustments be used for civil society, as Polanyi seems to suggest.

5

Starting with the assumption of a 'pure' civil society, without any shared convictions, there is obviously no common criterion of good or bad other than the preservation of Society itself. It follows that individuals may indulge their personal preferences just as far as this does not threaten to break or wear down the social linkage. Supposing myself a member of such a society, I shall find it advisable to humour the preferences of other members, whatever they may be, thus preserving good relations with them and with a view to reciprocity. The maxim of such conduct is properly designated as 'Live and let live'. Now there is a vast realm within which Ego and Alter, supposedly concerned with nothing other than their respective preferences, can, quite selfishly, by the process of bargaining, help each other to reach higher positions on their respective preference curves. The main novelty of modern civilization, as it has developed in the last two centuries, is the successful exploitation of compatible preferences. It does not matter to Ego that Alter wants something which Ego deems absurd, if this wanting of Alter spurs him to provide to Ego something which Ego wants.

Alter's passions are Ego's opportunity and vice versa. So far so good. But there are passions which either for circumstantial reasons or by virtue of their nature, do not lend themselves to mutually satisfactory horse-trading. When the case arises, the attitude of Ego to Alter changes most abruptly. At one moment Alter's preferences were Ego's opportunity; now they are a nuisance and a danger for Ego. An effort to assess fairly the moral value of Alter's attitude plays very little part in all this. Ego did not take the trouble to blame Alter's preferences when these were conducive to the satisfaction of Ego's preferences. Conversely, when Alter's preferences interfere seriously with the satisfaction of Ego's preferences, Ego does not take the trouble to see what can be said in justification of Alter's attitude. Easy-going tolerance, because it was based on subjective interest, pivots, on the same basis, to passionate intolerance.

When such an accident occurs, and if it involves important numbers, a problem arises for those who are in charge of the body politic. Some sacrifice of preferences has now to be decreed. But, *ex hypothesi*, the political managers have no other aim than to preserve Society in being; this affords them no criterion to assess which of the warring preferences are the best justified. Their only criterion being the preservation of the social state, it follows that what must guide them in the settlement is the objective of minimizing the anti-social feelings aroused by some sacrifice of preferences. This naturally leads them to lay whatever sacrifice of preferences is necessary to a settlement upon those who are least apt to react passionately against this settlement. Hence whenever things come to an open clash of preferences, the referees will bias their decision in favour of the most prone to unruliness, the propensity to turbulence thus becoming a political asset; in a society without decisive principles or standards it must logically be so.

Polanyi would certainly assert that in the society we are imagining, the lack of any basic set of values is by no means to be confused with a dearth of moral passions, as these are natural to man: it would not be the case that most or the main clashes of preferences would arise from conflicts of material interests: the most serious would be due to moral passions running riot. In the absence of any framework for these passions, such passions could be assessed on the basis solely of their intensity, not of their direction. According to the argument presented in the foregoing paragraph, political referees looking solely to the preservation of social cohesion would tend to 'appease' (in the discreditable meaning acquired by the word) the passions most fractiously manifested; moreover, there being no criterion by which moral feelings can be said to be well or ill directed, it might well seem that the very violence of a certain moral passion would testify in its favour: where no truth is acknowledged, fanaticism is impressive.

6

If we now turn to Polanyi's model of a dedicated company, quite another prospect unfolds. Here it is certainly not true that each member is welcome to pursue the satisfaction of his subjective preferences, whatever they may be. What brings and keeps these men together is the pursuit of knowledge. An individual scientist may lose his appetite for discovery: but thereby he forfeits his partnership. In this company the 'live and let live' maxim is not valid. Suppose that Primus is an acknowledged authority in a given field of science: he likes Secundus and wishes him well; therefore he praises a work of Secundus which he knows to be invalid; or again he does so for a far more discreditable reason, because Secundus can do him a favour. Now it does not matter whether such praise is inspired by mere kindness or by self-interest: the act itself is a major sin. The reason thereof is that 'nobody knows more than a tiny fragment of science well enough to judge its validity and value at first hand.'[8] It follows that the seal of Primus placed upon the work of Secundus will give it currency throughout the scientific community. Even in a distant compartment, it will be received by scientists who possibly have hardly heard of Primus, but who will have received the findings mediately through a continuous chain of authorities of which Primus forms the first link: it has to be a trustworthy link, or else spurious findings may be relied upon here and there and spoil the work done at many removes.

Consequently each scientist has to maintain rigorous standards, not only in his own labours, but in his judgement of the work of others, within the territory of which his qualifications make him the natural and necessary censor. The recognition he has gained entails a confidence placed in him which he may not betray. Of course it is pretty hard to condemn the work of a fellow scientist for whom one feels sympathy. It is something of a paradox, that the general attitude of vague and loose leniency which can so easily be maintained in civil society towards people who are indifferent to us, cannot be maintained in the scientific community where the links of personal sympathy are naturally stronger than in civil society. However mellowed by kindness, an attitude of mutual censorship is a requirement of a dedicated company: only at this price can scientists rely upon 'scientific work'. An amiable character will be troubled by withholding eagerly-awaited praise, even worse by passing condemnation. But individuals are incited to maintain their standards of rigour by an undevised system of cross-checks:

'Each scientist watches over an area comprising his own field and some adjoining strips of territory, over which neighbouring specialists can also form reliable first-hand judgements. Suppose now that work done on the speciality of B can be reliably judged by A and C; that of C by B and D; that of D by C and E; and so on. If then each of these groups of neighbours agrees in respect to their standards, then the standards on which A, B, and C agree will be the same on which B, C, and D agree, and on which also C, D, and E agree, and so on, throughout the whole realm of science. This mutual adjustment of standards occurs, of course, along a whole network of lines which offers a multitude of cross-checks for the adjustments made along each separate line; and the system is amply supplemented also by somewhat less certain judgements made by scientists directly on professionally more distant achievements of exceptional merit. Yet its operation continues to be based essentially on the 'transitiveness' of neighbouring appraisals—much as a marching column is kept in step by each man's keeping in step with those next to him.

'By this consensus, scientists form a continuous line—or rather a continuous network—of critics, whose scrutiny upholds the same minimum level of scientific value in all publications accredited by scientists. More than that: by a similar reliance on each immediate neighbour they even make sure that the distinction of scientific work above this minimum level, and right up to the highest degrees of excellence, is measured by equivalent standards throughout the various branches of science. The rightness of these comparative appreciations is vital to science, for they guide the distribution of men and subsidies between the different lines of study, and they determine, in particular, the crucial decisions by which recognition and assistance are granted to new departures in science or else withheld from them. Though it is admittedly easy to find instances in which this appreciation has proved mistaken, or at least sadly belated, we should acknowledge that we can speak of 'science' as a definite, and on the whole authoritative, body of knowledge only to the extent to which we believe that these decisions are predominantly correct.'[9]

7

The Republic of Science is not subject to a supreme central authority. But it is subject to a system of control which is of quite an oligarchic hue. If freedom of opinion is meant to imply the opportunity to express publicly whatever one would like to say, then there is no such freedom in the Republic of Science. The opportunity to make one's work known, or one's reaction to the work of another, is afforded by scientific journals, wherein scientific papers can be published only with the preliminary approval of two or three independent referees called in by the editor of the journal. The referees express an opinion particularly on two points: whether the claims of the paper are sufficiently well substantiated and whether it possesses a sufficient degree of scientific interest to be worth publishing. . . .[10]

'The referees advising scientific journals will also encourage to some extent those lines of research which they consider to be particularly promising, whilst discouraging other lines of which they have a poor opinion. The dominant powers in this respect, however, are exercized by referees advising on specific appointments, on the allocation of special subsidies and on the award of distinctions. Advice on these points, which often involve major issues of the policy of science, is usually asked from and tendered by a small number of senior scientists who are universally recognized as the most eminent in a particular branch. They are the chief Influentials, the unofficial governors of the scientific community. By their advice they can either delay or accelerate the growth of a new line of research. They can provide special subsidies for new lines of research at any moment. By the award of prizes and of other distinctions, they can invest a promising pioneer almost overnight with a position of authority and independence. More slowly, but no less effectively, a new development can be stimulated by the policy pursued by the Influentials in advising on new appointments. Within a decade or so a new school of thought can be established by the selection of appropriate candidates for Chairs which have fallen vacant during that period. The same end can be advanced even more effectively by the setting up of new Chairs.'[11]

At this point the picture arouses some disquiet: so much seems to rest upon that 'small number of senior scientists' who apparently control all avenues. Polanyi does not call attention, however, to the dangers of this structure of power, but to its efficiency. Familiar as he is with the world of science, he would, of course, denounce such evils if there were occasion to do so. But this leaves us with a question-mark. Is it not strange that the internal constitution of the Republic of Science should present traits so similar to those which scientists ardently denounce in the body politic whenever they utter political opinions?

7

The organization of the community of scientists has an obvious bearing upon the problem of political organization. When Polanyi describes the training of a future scientist, the education which incorporates scientific standards into his way of thinking and of feeling, this cannot fail to suggest the education which youths received in the Greek *polis* and which was therein regarded as indispensable to the making of a citizen. When he speaks of 'censorship' it is obviously not in the modern sense of shutting people's mouths but with the meaning attached to the word in the Greek and Roman Republics, where censoring meant putting people in the place which they deserved, or down-grading them if their conduct called for this. Thus the Roman censor excluded from the Senate those who had not performed adequately: censoring meant, and means to Polanyi, keeping citizens up to the mark, 'maintaining the standards of performance'. The prompt and severe retribution awaiting the scientists who have published spurious results is reminiscent of the dire punishment meted out to the Greek citizen who had abused his right of speech to formulate a proposition later found out to have been ill-advised. The lack of any centralized authority and the remedy for this in the guiding role of the Influentials again form a trait characteristic of the ancient republics. But most decisive of all likenesses is the fact that mere presence does not make a citizen: it requires participation in the common cult.

Large, sprawling, diverse modern society is definitely not a *polis*. It is perhaps in order to note that the Catholic Church, when it ceased to be composed of small churches of ardent believers and came to deal with all, 'tempered the rigour of principles by the mildness of discretion,' in the words of Gregory VII. It is a simple idea but an important one that the discipline of the dedicated cannot be extended to the many without hateful despotism. That, indeed, is at no point suggested by Polanyi. Surely it can be held that the body politic differs fundamentally from the Republic of Science in that the latter has a vocation while Society has none. But Polanyi would clearly be unwilling to adopt the latter statement, he does not want to think of Society as an undedicated gathering, content to benefit from the findings of scientists. To him it also must be a dedicated body. Dedicated to what? To 'continuously reshaping its own life in the pursuit of civic virtue freely fostered in its midst.' (^{12}). If so, just as the body scientific is a 'conspiration' for the dynamic unfolding of truth, the body politic would be a 'conspiration' for the dynamic unfolding of virtue. But then, having stressed the conditions which are germane to the pursuit of the first purpose, Polanyi would lead up to the question of the conditions germane to the second purpose. Whether or not one subscribes to the latter, the description of the community of scientists is richly suggestive.

NOTES

1. These Riddell Memorial Lectures were published in 1946 under the title *Science, Faith and Society*.
2. *The Logic of Liberty*, 1951, pp. 53 et seq.
3. *P.K.*, 1958, e.g. pp. 163, 217, etc.
4. The subject is dealt with in some thirty pages of S.F.S., pp. 29–51, where a great deal is to be found which goes beyond the subject as delimited here.
5. *S.F.S.*, p. 29.
6. *S.F.S.*, pp. 29–30.
7. *S.F.S.*, p. 50.
8. *P.K.*, p. 163.
9. *P.K.*, pp. 217–18.
10. *L.L.*, p. 53.
11. *L.L.*, p. 54.
12. *P.K.* p. 223.

13

Machiavelli and The Profanation of Politics

IRVING KRISTOL

THE *Secretum Secretorum* is a brief treatise attributed to Roger Bacon that had great currency during the later Middle Ages, in various 'editions' and under various forms of the title *De Regimine Principum* as well as its own. It presents itself as a letter of advice from Aristotle to his student, Alexander of Macedon, who was having trouble ruling the Persians he had just conquered. In the course of this letter, Aristotle says that a king should put God's law before his own; avoid the sin of pride from which all other sins flow; converse with wise men; help the poor and needy; flee from lechery and lust; never break his oath; enjoy music while remaining of grave countenance; etc., etc. By far the most interesting thing about the *Secretum Secretorum* for the modern reader is its title. What, one wonders, is the secret?

The answer appears to be, quite simply, that there is none. Not really. All the title signifies is that the art of good government is something so rare, something which so few men ever discover, that it can be considered a hidden treasure. I have said 'the art of good government'—the art of *self-*government would have been more precise. For the whole vast medieval and early modern literature concerned with *De regimine principum, De officio regis, De institutione principum,* and so on—Professor Allen Gilbert's *Machiavelli's 'Prince' and its Forerunners* gives us an idea of the number and scope of such works—intends primarily to instruct rulers on how to govern *themselves*. This is, under any conditions and for any man, the most difficult task in the world. It is especially difficult for a man who, like the prince, is surrounded by the temptations that go with wealth, power, and an atmosphere of servile flattery. In most cases, of course, since princes are only human, the effort is bound to fail. Such failures are not described in this highly moralistic literature, in part doubtless for fear of undermining the authority of government itself, but mainly for fear of setting a bad example.

The fact that this political literature was so little 'sociological', so blandly mindless of economics, administration, even the military arts, obviously reflects to some extent the simple conditions of medieval life and medieval society. But only to some extent. If it is true that what the king did was less important than what kind of king he was; and if this in turn was less important than the fact that he was, indisputably, the rightful king—nevertheless, he had much to learn about his world that these guides never attempted to teach. Their self-limitation and, to our eyes, curious modesty appears to derive from their assumption that whereas morals, involving as it does a knowledge of the good, may be improved through exhortation and instruction, wise government is a practical activity that cannot be divorced from specific circumstances and which therefore can only be 'learned' through the experience of ruling.

'Political science' in the medieval sense meant the description of obligations; it gave no further practical advice for it did not claim any special practical wisdom. It did not deny the existence of such wisdom; it simply denied that philosophers, as against statesmen, possessed it. It is one thing to say a king should be merciful; it is quite another to say that he should spare the life of a particular conspirator—for who can foresee whether such an individual act of mercy might not mean the ruin of the commonwealth and general misery? Providence, of course, knew the answer. But Providence was inscrutable, as much to philosophers as to anyone else. Only prophets could read the future; and all (including the Church) agreed that the age of prophecy was over. For a philosopher to attempt to judge human action by its consequences, instead of its concord with the moral law, was to claim a superhuman ability to foresee the general and ultimate results of specific actions. It was as a denial of this human ability that the Ten Commandments, and the moral code associated with it, were proclaimed as authoritative. And it was because he believed that men did not—and in the nature of things could not—have such power that St. Thomas said flatly: '*Eventus sequens non facit actum malum qui erat bonus, nec bonum qui erat malus.*'

It is against such a background that one can appreciate the revolution in political theory that Niccolò Machiavelli accomplished. To be sure, the older order of thought did not vanish overnight. The year in which *The Prince* was probably finished was also the year of two such popular works as Erasmus's *Education of a Christian Prince*, an eloquent homily, and Thomas More's *Utopia*. This latter was in the classical rather than in the medieval tradition, but these had more in common with each other than with the modern mode. More's Utopia was located, not in the future, but out of time entirely; it posed an ideal and criticized reality in its name—but it did not suggest that reality could be transformed into ideality through political action. It was a purely normative exercise. Within the book itself, More inserts a dialogue on what role philosophers can play in politics (he had just been offered a post by Henry VIII), and concludes that, at best, he can by his counsel prevent some evil from being done. This was hardly what we would today call utopian doctrine. And, in the event, More's own martyrdom was to reveal that even this 'at best' was an elusive possibility.

The homiletic tradition, then, continued after Machiavelli. Indeed, one finds innumerable specimens of the genre in the sixteenth and seventeenth centuries. But one also finds that, under the influence of 'Machiavellianism', this genre is either being converted into, or tinged with, something new in political philosophy. This 'something new' lies not merely in the fact that Machiavelli stands the tradition upon its head. He does do that. Whereas it had claimed moral authority and disclaimed political knowledge, he repudiates the established moral authority and asserts a kind of knowledge that the tradition did not recognize. Yet there had been Christian sects which insisted that the moral law had to be abrogated so as to prepare the way for the Second Coming; and there had been sects, too, which felt that, as a result of some secret communion with either God or the devil, they had been supplied with 'the key' to man's temporal destinies. Machiavelli is no Christian heretic; he is the first of the post-Christian philosophers.

Post-Christian, not pre-Christian. Since Machiavelli lived during the Renaissance and, like all Renaissance writers, continually referred to Greek and Roman authors as his authorities, one is inevitably tempted to see in his thought a resurgent paganism. But a careful reading makes it obvious that Machiavelli uses his classical 'authorities' in an arbitrary—and often downright cynical—manner. Moreover, the very spirit that pervades Machiavelli is markedly different from that which finds expression in, say, Thucydides or Tacitus. The classical writers, like Machiavelli, had no conception of Providence, believed that men were the toys of Chance and Necessity, and admitted that the universe was blind to human values—but they also asserted (or, at the very least, implied) that man could be superior to his fate in so far as he faced it with nobility of character, courage, and grace. Their writings breathe a *pietas* before the cosmic condition of the race; whereas Machiavelli writes with the sardonic iciness of inhuman fate itself. He is the first of the nihilists, not the last of the pagans.

This is not to suggest that he was devoid of human feeling. His passionate Italian patriotism was, for instance, doubtless genuine enough. So was the streak of sadism (no other word will do) that runs through his work, from his very first opuscule to his last. But these sentiments, though they have an important effect on the literary quality of his work, and help explain both its popularity in Italy and its notoriety everywhere, are subsidiary to his main purpose. This purpose, as announced in *The Prince*, is to describe 'the way things really are' (*la verità effettuale della cosa*) rather than—as the medieval theorists had done—the 'imagination' of it. 'For many have pictured republics and principalities which in fact have never been known or seen, because how one lives is so far distant from how one ought to live, that he who neglects what is done for what ought to be done, sooner effects his ruin than his preservation.'

But what, a modern reader is bound to ask, was so shocking about that? What, moreover, is 'nihilistic' about it? Is it not a sensible attitude—indeed a 'scientific' attitude?

These questions are best answered by another question: what in our own time is so shocking about de Sade? We know that the kinds of sexual activities he describes do exist and play an important role in men's lives. Lust, adultery, sodomy, pederasty, and all the various sexual aberrations have always been with us; and there is no question that they are necessary to the 'happiness' of a large class of people. Yet in no country of the world may de Sade's books circulate freely. Our society seems to believe that unrestricted knowledge of these subjects constitutes pornography. It insists that, if they are to be discussed at all, it must either be in an esoteric manner (in medical textbooks) or within a moral framework that makes it clear one is treating of an evil, not merely a human phenomenon. And, for de Sade, there is no natural and prescriptive moral framework in sex, just as there is none in politics for Machiavelli.

Pornography may be defined as a kind of knowledge which has an inherent tendency to corrupt and deprave our imaginations. The twentieth century formally recognizes that pornography, as such, does exist; but it also feels committed to the contradictory thesis that knowledge *per se* is good and 'enlightening'. This is but another way of saying that the twentieth century is experiencing a crisis of values—not simply a conflict between values, but a crisis in the very idea of value. For if one allows that knowledge in and of itself may be the supreme value, one must go on to say that the knowledge of evil is as valuable as the knowledge of good, from which it flows that a man who is engaged in adding to our knowledge of evil is as virtuous as a man engaged in adding to our knowledge of good—in short, that the difference between evil and good is at most a matter of habitual terminology. This is, precisely, nihilism.

One cannot appreciate the new *frisson* that Machiavelli gave to his age without realizing that he appeared to his contemporaries as a kind of political pornographer. 'I hold there is no sin but ignorance,' Marlowe has him say in one of his plays. The ascription was entirely apt. For the message of Machiavelli was really nothing more than the message of pornographers everywhere and at all times: that there is no such thing as pornography. Nothing that Machiavelli said about affairs of state was really novel to his readers. They knew—everyone had always known—that politics is a dirty business; that a ruler may better secure his power by slaughtering innocents, breaking his solemn oaths, betraying his friends, than by not doing so. But they also knew, or had thought, or had said, that such a ruler would suffer the torments of hell for eternity. Where Machiavelli was original was first, in brazenly announcing these truths, and second, in implying as strongly as he could (he dared not be candid on this subject, for it would have cost him his head) that wicked princes did not rot in hell for the sufficient reason that no such place existed.

These two aspects of Machiavelli's originality were most intimately connected, and necessarily so. Had he accepted Christian morality and the prospect of divine judgement, he would never have wanted to break the traditional silence on the awful things men do in their lust for power; he would have been fearful of depraving the imagination of men, especially of princes, and of incurring responsibility for their damnation. Had he even accepted the moral code of the Graeco-Roman writers (who did not believe in divine judgment either) he would at least have indicated how awful these things were, no matter how inevitable in the course of human affairs. But instead he declared that an honest and enlightened man had no right to regard them as awful at all. They were inherent in the nature of things; and with the nature of things only fools and sentimentalists would quarrel. The classical writers knew that the rule of tyrants was an intrinsic possibility of human politics—one that was bound to find realization under certain circumstances that made tyranny the only alternative to chaos or foreign domination. They might 'justify' tyranny; but without ever denying that it was tyranny. Machiavelli, in contrast, wrote a book of advice to aspiring tyrants in which the word 'tyranny' simply does not appear.

There is in Machiavelli a deliberate, if sometimes artful debasement of political virtues. One of the secrets of his sparkling style is the playful way he gravely uses the conventional rhetoric in order to mock its conventional character. Thus, in his eulogy of Castruccio Castracani he writes solemnly: 'He was just to his subjects, faithless to foreigners, and he never sought to conquer by force when he could do so by fraud.' Such examples can be multiplied a hundredfold. His constant use of the term '*virtù*' to mean that which characterizes the virtuoso is perhaps his outstanding pun (e.g., Agathocles 'accompanied his infamies with so much *virtu* that he rose to be praetor of Syracuse.') In the course of undermining the traditional political virtues, he also takes the opportunity—wherever possible—to show contempt for the established religion. He can do so under the guise of 'interpreting' Biblical history in his *Discourses*, as when he solemnly praises Moses and King David for their cruelty and ruthlessness; or he can do so more openly in a 'technical' work like the *Art of War*, where he blames Christianity for the decline of martial prowess in Italy.

But the most candid statement of 'Machiavellianism' is in his *Florentine History*. It is put into the mouth of 'one of the boldest and most experienced' of the plebeian leaders during the revolt of 1378; but there can be no doubt that it is Machiavelli himself who is speaking from the heart:

'If we had now to decide whether we should take up arms, burn and pillage the houses of the citizens, and rob the churches, I should be the first among you to suggest caution, and perhaps to approve of your preference for humble poverty rather than risking all on the chance of a gain. But as you have already had recourse to arms, and have committed much havoc, it appears to me the point you have now to consider is, not how shall we desist from this destruction, but how we shall commit more in order to secure ourselves. . . . It is necessary to commit new offences by multiplying the plunderings and burnings and redoubling the disturbances . . . because, where small faults are chastised, great crimes are rewarded. . . . It grieves me to hear that some of you repent for consciences' sake of what you have already done and wish to go no further with us. If this be true you are not the sort of men I thought you were, for neither conscience nor shame ought to have any influence upon you. Remember that those men who conquer never incur any reproach. . . . If you watch the ways of men, you will see that those who obtain great wealth and power do so either by force or fraud, and having got them they conceal under some honest name the foulness of their deeds. Whilst those who through lack of wisdom, or from simplicity, do not employ these methods are always stifled in slavery or poverty. Faithful slaves always remain slaves, and good men are always poor men. Men will never escape from slavery unless they are unfaithful and bold, nor from poverty unless they are rapacious and fraudulent, because both God and Nature have placed the fortunes of men in such a position that they are reached rather by robbery than industry, and by evil rather than by honest skill. . . .'

This is strong medicine indeed, and it was precisely as a kind of strong medicine that Machiavelli was first apologetically presented to the world. When his *Prince* was posthumously published in 1532, the printer, Bernardo di Giunta, dedicated it to Monsignor Giovanni Gaddi, asking to be protected from those critics 'who do not realize that whatever teaches of herbs and medicines, must also teach of poisons—only thus can we know how to identify them'. This medical metaphor has been fairly popular with writers on Machiavelli ever since (e.g. Ranke and Macaulay). It testifies to a recognition that Machiavelli *can* be a dangerous teacher; but it also claims that he may be a useful one.

Useful for what? To this question, there have been many answers, and a summary of them would be nothing short of a history of Machiavelli's influence on modern thought—it might be nothing short of a history of modern thought itself. But four of these answers are most prominent and most popular.

1. *The historical-scholarly answer.* The scholarship on Machiavelli and his times has been voluminous, technically superb, and almost invariably misleading. The bulk of this work has been done by Germans and Italians, and in both these countries the growing interest in Machiavelli was concurrent with efforts to form a united nation. For a century and a half after *The Prince* appeared, the commentators on Machiavelli—whether friendly or hostile—paid not the slightest attention to the final chapter, with its exhortation to free Italy from the barbarians. It was Herder who first saw in this the key to Machiavelli's thought, and who set the tone for modern scholarship. The tendency of this scholarship is to admire Machiavelli as one of the ideological founders of the modern national state, and it has seen in his seeming amorality a gesture of desperate patriotism and bitter

Machiavelli and the Profanation of Politics

pathos, suitable to his corrupt epoch. British scholars (usually Italophile) have also been inclined to follow this interpretation. This explains how it is that Macaulay came to say of Machiavelli's *oeuvre* that 'we are acquainted with few writings which exhibit so much elevation of sentiment, so pure and warm a zeal for the public good, or so just a view of the duties and rights of citizens', or that in our own day T. S. Eliot could assert, 'such a view of life as Machiavelli's implies a state of the soul which may be called a state of innocence.'

Now, Machiavelli was certainly an Italian patriot. But (as Professor Leo Strauss has demonstrated, in what is by far the best book on Machiavelli yet written) he was a patriot of a special kind. 'I love my country more than my soul,' he wrote to Guicciardini; and that he was sincere may be gathered from those scattered remarks in the *Discourses* where he emphasizes that, when a nation's interests are involved, no considerations of justice, legality, or propriety ought to affect our judgement. Whether one finds this laudable or not will, of course, depend on the relative estimates one places upon one's fatherland and one's soul. Very few of the scholars who admire Machiavelli are explicit on this point. A few, following Friederich Meinecke, concede resignedly that it is the ineluctable nature of political life to lead patriotic souls to perdition; though after the German experience of the past thirty years, one may expect to hear less of this. But, in any event, the basic trend of conventional Machiavelli scholarship is to suggest to the student that if a man cares dearly for his country, it does not much matter what else he cares for.

2. *The* raison d'État *answer*. It is reported that Mussolini kept a copy of *The Prince* on his night-table. For all the good it did him, he was following an old tradition that goes back to the sixteenth and seventeenth centuries, when kings and ministers surreptitiously read Machiavelli or pale imitations of him in order to glean the esoteric and dreadful wisdom of *raison d'État*. (After Machiavelli was condemned by the Church, they may have shifted to Tacitus, who during that period was taken to be a proto-Machiavellian.) For with the rise of the absolute monarchies, there was a need for a theory of the State. The previous political theory, not of the State but of Society—the theory of the Christian commonwealth, in which kingship was a well-defined office—had been rendered archaic; and into the vacuum thus created there rushed the esoteric doctrine of 'reason of state'. What this doctrine came down to was that (a) it was perfectly legitimate for a king to extend or secure his power and dominion by any and every means, i.e. to act like a tyrant; and (b) his subjects must be left in ignorance of this truth lest it undermine their pious subservience to what passed for 'duly constituted' authority—the king had to be hypocritical as well as unscrupulous.

This whole historical episode, during which the fashion of dabbling in *raison d'État* was the rage of courtiers, ministers, confessors, and paramours, has not yet been adequately told. The few ponderous German studies of it, properly humble before something that has the air of *Realpolitik*, completely miss its farcical aspect. For the 'rules' of *raison d'État* are very similar to—they are sometimes identical with—the familiar household proverbs that can be quoted to suit any purpose. ('Look before you leap,' and 'He who hesitates is lost,' etc.) Machiavelli is full of general rules and prescriptions—all of which conflict with one another, and some of which, as Professor Butterfield has shown, are patently contradicted by the evidence he marshals for their support. Such a state of affairs is unavoidable, since generalizations of this order have no purchase upon experience. When a king should murder his defeated enemies, and when he should treat them leniently, is not something that can be decided *a priori*—it is difficult enough to decide it *a posteriori*, as historians know. The statesman who tries to substitute abstract deductions for prudent judgement is not long for this world.

In fact, the rulers of the sixteenth and seventeenth centuries managed to survive reasonably well, and the most clever and resourceful of them prospered mightily. This was not because of anything they learned from their readings in the new 'philosophy' of politics. They did what they thought was the sensible thing to do under the circumstances; and all that '*raison d'État*' constituted was the reassurance that whatever they did need not trouble their consciences. This perhaps made them a little more brutal than they might otherwise have been; but one can never be sure. As Machiavelli himself said, rulers had long practised what he first preached.

3. *The democratic-enlightenment answer.* This has been by far the most influential of all, and it derives directly from the medical metaphor proposed by Bernardo di Giunta. Machiavelli is taken for an acute anatomist and diagnostician of political disorder, who has exposed the unscrupulousness of rulers in order to allow men to recover their political health in pure self-government, i.e. popular government.

In its most extreme form, this view regards Machiavelli as a cunning satirist, and his *Prince* as a Swiftian, self-defeating 'modest proposal'. Though no less an authority on the Renaissance than Garrett Mattingly has recently restated this thesis, it is no more persuasive today than when it was first suggested by Alberico Gentile at the end of the sixteenth century. It involves, to begin with, a reading of the *Discourses* as a 'republican' document that expresses Machiavelli's true convictions. Yet, as Macaulay pointed out in rejecting this possibility, all the 'Machiavellian' sentiments of *The Prince* are also to be found scattered through the *Discourses*. There is also the fact that when *The Prince* circulated in manuscript before Machiavelli's death in 1527, considerable odium was attached to it by the Florentine republicans, who saw it as a pro-Medici tract. Machiavelli, as we know from his play, *Mandragola*, was capable of first-rate satire; it is implausible that he would have so botched the job in *The Prince* as to make it produce an opposite effect to what was intended.

Machiavelli and the Profanation of Politics

The main current of thought which takes Machiavelli as a precursor of 'enlightenment' is content to see in him merely an honest man who exposed the trickery of princes. Trajano Boccalini, in his *News from Parnassus* (1612), recounts a tale in which Machiavelli, having been banished from Parnassus on pain of death, was found hidden in a friend's library. Before the court of Apollo, he enters the following plea in his self-defence:

'Lo hear, you Sovereign of Learning, this Nicolas Machiavel, who has been condemned for a seducer and corrupter of mankind, and for a disperser of scandalous political precepts. I intend not to defend my writings, I publicly accuse them, and condemn them as wicked and execrable documents for the government of a State. So if that which I have printed be a doctrine invented by me, or be any new precepts, I desire that the sentence given against me by the judges be put in execution. But if my writings contain nothing but such political precepts, such rules of State, as I have taken out of the actions of Princes, which (if your Majesty gives me leave) I am ready to name, whose lives are nothing but the doing and saying of evil things—then what reason is there that they who have invented the desperate policies described by me should be held for holy, and that I who am only the publisher of them should be esteemed a knave and an atheist? For I see not why an original should be held holy and the copy burnt as execrable. Nor do I see why I should be persecuted if the reading of history (which is not only permitted but is commended by all men) has the special virtue of turning as many as do read with a politic eye into so many Machiavels: for people are not so simple as many believe them to be (and have) the judgement to discover the true causes of all Prince's actions, though they be cleverly concealed."

The judges are so impressed by this logic that they are ready to release him, when the prosecuting attorney reminds them of their responsibility: 'For he has been found by night amongst a flock of sheep whom he taught to put wolves' teeth in their mouths, thereby threatening the utter ruin of all shepherds....' And for this, Machaivelli is duly burnt on Olympus.

He fared much better, however, down on earth, where shepherds were beginning to lose their good repute, as a preliminary to losing their heads. Harrington saw in Machiavelli the Hippocrates of the body politic; Spinoza praised him by name (a rare honour) as *prudentissimo*; Diderot flattered him in his Encyclopedia; while Rousseau eulogized *The Prince* as '*le livre des republicains*' and its author as one who 'pretending to give lessons to kings, gave some important ones to the people'. Even John Adams admired him as a republican benefactor.

It is easy to see how Machiavelli's work of 'enlightenment' suited the various thinkers of the Enlightenment. Their project was the discrediting of traditional political authority and the revelation to all of the *arcana imperii*, so that the rule of special privilege could be replaced by the sovereignty of the common good.* Machiavelli was all the more attractive in that his writings do contain several laudatory references to popular government, which seemed to give him a 'democratic' bias. This was, of

---

* For the way in which this moral passion was inverted into a set of fanatical ideologies, see Michael Polanyi's Eddington lecture, *Beyond Nihilism*, Cambridge University Press, 1960 (and *Encounter* March 1960), as well as *P.K.* ch. 5.

course, a misreading. Tyranny and democracy were not, for Machiavelli; exclusive conceptions; and his notion of popular government was sufficiently elastic to include the kind of rule projected by the popular leader of 1378, in the speech already quoted. But the men of the Enlightenment were not much worried about the future of popular morals; they took the moral instinct as natural, unless corrupted by government, and foresaw the progressive accommodation of human government to innate human goodness. The best state was the one that made its own existence as near to superfluous as possible; and any literature which cast obloquy on the medieval idea of the state as a coercive force necessary for man's mundane perfection, was welcome.

4. *The 'positivist' answer.* Like the nationalist answer, this is of more symbolic than practical significance, since it involves only the corruption of professors. It belongs to the twentieth century and most particularly to America; though it was first stated by Francis Bacon ('We are much beholden to Machiavel and others that wrote what men do, and not what they ought to do'), was revived for our time by Sir Frederic Pollock, and is now being promoted in Europe together with the rest of American 'political science'. According to this view, Machiavelli was a predecessor to Professor Harold Lasswell in trying to formulate an 'objective' set of political generalizations derived from, and to be tested by, experience. His seeming amorality is nothing but the passionless curiosity of the scientific imagination.

It is obvious that this interpretation is incompatible with the medical metaphor, and with the idea that the political thinker is a physician to the state. Medicine, after all, is a normative and practical discipline, in that it has an ideal of bodily health to which its activities are subordinated. Even medicine's allied sciences (anatomy, physiology, etc.) share this character: structure is studied in terms of function, function in terms of structure, and the whole is related to an ideal human organism—'ideal' in the Aristotelian sense of most appropriately 'according to nature'. The 'positivist' approach—I use inverted commas since the term itself is a source of contention—refers to physics as its model instead of to medicine. It proposes to establish demonstrative 'truths' about men in politics that will be available to whatever set of 'values' wishes to employ them.

Were this line of thought as fruitful as its proponents think it might be, it would itself pose a major political problem. No government could allow such potent truths to enter freely into political life—any more than it can permit the knowledge of how to make atom bombs to circulate freely. Political scientists who were not content to stick to general theory and academic publications, and who tried to apply their knowledge to specific problems, would have to obtain a security clearance and work under official supervision. Sometimes one gets the impression that the political scientist, in his envy of the intellectual authority of the physical scientist, would not in the least mind such flattering coercion. But, fortunately, the 'demonstrable truths' of political science have so far been relatively trivial. And there are even many who think the whole enterprise is misconceived—that it is as senseless for 'political scientists' to try to achieve an 'objectivity' towards political man as it is for medical science to seek such objectivity towards the human body.

It is interesting, nevertheless, that the assertion should be made—that an influential and reputable group of scholars should insist that it is *right* for political knowledge to be divorced from moral knowledge. This goes a long step beyond the older *raison d'État*, which merely recognized, and took advantage of, their frequent incongruence. Machiavelli would have approved; though he would have been properly sceptical of the willingness of academic persons to carry this assertion through to its boldest implications.

There have been three major figures in the history of Western thought during the last five centuries who have rejected Christianity, not for its failure to live up to its values, but because they repudiated these values themselves. The three are Machiavelli, de Sade, and Nietzsche. A great part of the intellectual history of the modern era can be told in terms of the efforts of a civilization still Christian, to come to terms with Machiavelli in politics, de Sade in sex, Nietzsche in philosophy. These efforts have been ingenious, but hardly successful. The 'slave morality' of Christianity is constantly in retreat before the revolt of 'the masters', with every new *modus vivendi* an unstable armistice. Heidegger has even gone so far as to say that the struggle is over—that with Nietzsche the Christian epoch draws to a close. If this is so, then it can also be said that Machiavelli marks the beginning of this end.

14

Applied Economics— The Application of What?*

ELY DEVONS

ALTHOUGH I have been an 'applied economist' for many years and have frequently tried to be introspective about my activities, I am, I fear, not yet able to give a clear and methodical account of what it is I am applying. My readers will have to be satisfied, therefore, with random, not very clearly formulated thoughts, almost personal confessions, rather than a carefully worked out systematic treatment.

The main questions I would like to answer are, 'what is it one applies in applied economics?'; 'what is the nature of the understanding of reality one gets through this application?'; 'what use, if any, is this understanding in the formulation of policy?', and lastly, 'what are the best ways of teaching this applied economics?'. As I have already indicated, I cannot give satisfactory answers to these questions, but I hope that you agree that they are interesting and important questions.

I am going to divide my argument into three sections. In the first I shall discuss the uses of theory, in the second the other kinds of knowledge which are apparently used in analysing and discussing real economic problems, and in the third I shall discuss teaching.

First then, theory and its uses. I cannot, of course, pretend to cover the whole corpus of theory and all possible applications. I shall merely attempt to expound and support my general thesis by illustration and example, although I hope that some of these examples are concerned with the most relevant and important parts of theory. Perhaps it will be as well if I were to state my general thesis at the outset. This is, that in so far as economic theory is useful in enabling us to understand the real world and in helping us to take decisions on policy, it is the simple, most elementary and in some ways most obvious propositions that matter. I do not want, at this

* This is a paper read by Professor Devons at the A.U.T.E. conference at Southampton in December 1958.

stage, to argue whether this is merely a reflection of the present state of economic theory, or whether it is necessarily true of economic theory as such, although this may come up in discussion. I must also emphasize that what I am putting forward is very much a personal view of what theory conveys to me, and you may well think it egotistical arrogance on my part to suggest that such a view is in any way representative.

In putting forward my argument about theory, I want to make a distinction between theoretical models, common-sense axioms, and theoretical concepts.

Two of the most important sets of theoretical models are those of a price system and those of the relation between income, production, employment, and expenditure. In both of these it is the elementary propositions conveyed by the models that I find relevant and usable.

Let me take models of the price system first. These are of two kinds. The general which portray the economy as a whole, and the partial relating to a single market or to closely interrelated markets. The general models, even of the most elaborate kind, serve the simple purpose of demonstrating the interconnectedness of all economic phenomena, and show how under certain conditions price may act as a guiding link between them. Looked at in another way, such models show how a complex set of interrelations can hang together consistently without any central administrative direction. They show prices, and the reaction of supply and demand to prices, as the great instruments of co-ordination between dispersed and decentralized decisions. Such models, therefore, give us an understanding of the general principles of operation of the economic system.

These general models are also frequently used as a picture of the perfect economic system, and therefore as providing criteria by which to decide how the real world ought to be made to behave. Such uses of the models are, however, full of pitfalls. For the models are usually static, and the criteria of efficiency and perfection used are themselves highly abstract and axiomatic. Even if the models can be accepted as relevant, it is dangerous to assume that the real world can be made to conform more closely to the model by policy manœuvres. Real life has other, often more important, aspects apart from the economic, and Government action, which is usually what is prescribed, has many side effects of major importance.

In my view when we get to such problems of applied economics as distribution of industry or monopoly policy, theoretical models of the price system help us little. It is true that argument often proceeds from comparisons between reality and models of perfection. Reality is then easily shown as imperfect, but it is too readily and glibly assumed that an alternative arrangement which it is proposed to put into operation will work perfectly. The real issue of a comparison between the known imperfections of present arrangements and the imperfections of potential alternatives, is one rarely made, and is one in which economic theory can be of little assistance. It is hardly surprising that in such fields of policy there are widely divergent views among economists.

Applied Economics: The Application of What?

It is in the second, the partial, Marshallian analysis, that I find price theory most useful. The laws of supply and demand, showing price as the mechanism for balancing supply and demand in particular markets, have extensive applications. And it is in their simplest form that they are most useful. Nothing in the way of complicated theory is involved. Let me take two examples to illustrate my view. First, Government or other intervention to fix prices in times of shortage during war or peace. With the simple model of market price as the mechanism for balancing supply and demand, it seems obvious to me as an economist that if the Government decides to fix a price below that warranted by demand, then some demand which would be prepared to pay a higher price must go unsatisfied, Some arrangement must, therefore, emerge in practice for deciding which demand will go unsatisfied. If these arrangements are left, so to speak to 'natural forces', the situation will be resolved by queues, under-the-counter dealing, allocation on the basis of friendship or influence, or by breaking the price regulation through a 'black market'. Hence the conclusion that if we do not like such arrangements, the system of price fixing must be accompanied by some enforceable system of rationing. I say enforceable, for unless the Government or other controlling authority can make sure that all supplies will be directed towards the ration, the price control will continue to be frustrated, to a greater or lesser degree, by a black market. From this elementary principle of the 'law of supply and demand', therefore, arises the whole complex of administrative measures that must be taken and the conditions that must be satisfied, if for some reason it is thought desirable to fix prices below the 'market' level. Similarly one can consider the relation with supply, and see the effect on supply of any intervention to fix the price above or below that which would result from the market-clearing principle of price.

My second example I take from University salaries. As you know, in the post-war period, the Treasury and the U.G.C., reinforced by pressure from the A.U.T. and the spirit of the age, have insisted on uniform salaries for the same grades of staff in all subjects. An Assistant Lecturer, whether in History, English, Classics, Physics, or Economic Statistics, must be paid the same salary. When I, as an economist, consider this insistence on uniformity, I see it in relation to the relative conditions of supply and demand of staff in these different subjects. And I come to the conclusion that if these conditions are different for different subjects, then the formal insistence on uniformity will have some peculiar consequences. The exact consequences I cannot tell in advance, but the kind of thing I expect might happen, and am not surprised by, is along the following lines. Suppose the salary is fixed at a level which merely calls forth an adequate supply in the popular Arts subjects, then it is unlikely to call forth an adequate supply in such subjects as Engineering, Physics, Accounting, or Economic Statistics. Since Assistant Lecturers will be difficult to obtain in these subjects, the Departments concerned will try, if the University system will let them, to appoint at the Lecturer level instead. If they get permission to do this, the net result will be that they will appoint as Lecturers staff of an age and with post-graduate experience comparable with Assistant Lecturers in such subjects as Languages and History where supply is more plentiful. Alternatively, if the general level of Assistant Lecturer salaries is such as to attract an adequate supply in the 'scarce' subjects, then in the 'Arts' subjects the supply will be superabundant, and the best candidates getting the Assistant Lecturer posts may have ten years or more research experience after graduation.

The point I am trying to make in these examples—and many others of a similar kind could, of course, be given—is that the economic theory used in analysing the situation for policy purposes is of the most elementary kind, very little beyond what is contained in the first chapters of Henderson's 'Supply and Demand'. But do not mistake my point, for I am not arguing that the understanding or results obtained from the application of the theory are themselves trivial. On the contrary, I think they are most valuable and important, and ignorance of them can lead to confusion and chaos in policy making. Nor does it follow from the simplicity of the propositions that they are easy to teach others to absorb into their way of thinking. Again it is characteristic of the situation that while an economist would approach the two sets of problems in the way I have explained, this approach would not normally occur to the administrator in the Civil Service or to the politician, or to non-economists in the University world. And even though some of them, and I emphasize *some*, might appreciate the point when put to them, they are unlikely to absorb it in such a way as to apply the principle, simple though it may be, to new situations.

The second group of models I want to refer to, to support my argument, are those which are used in discussion and decision about policy affecting the general level of economic activity. Here again I feel that however elaborate the theoretical models, econometric and other, it is the most simple and elementary elements in them which are used and are usable. Little more theory, to my mind, is used in policy making than the notion that income, production, employment, and expenditure are all closely interrelated, and that in maintaining employment and production it is 'effective demand' that matters. That if there are forces in operation making for a decline in production and employment, the way to restore them is by increasing effective demand, and vice versa.

We have perhaps got a little beyond this in dividing expenditure into what we regard as significant categories, and we use these divisions both in an attempt to analyse the determinants of fluctuations in expenditure, and in deciding how to influence these. We look separately at Government expenditure, construction, plant and equipment, stocks, consumption, and exports. But we have little usable knowledge in terms of general propositions about what determines fluctuations in expenditure in each of these divisions. All that theory seems to tell me at present is that I ought to watch each of these separately since they are subject to different influences, and changes in them have differing consequences. A good deal of the argument about economic policy results from differing views about what is happening or likely to happen to any one or all of these, and in relation to any particular diagnosis, from differences of view about the relative effectiveness and social and political desirability of particular policies for influencing them.

I know that there are also great controversies about the clash between domestic and foreign trade policies, about the risks that should be run in having unemployment, about the advantages or disadvantages of a rising price level, and the effect of these on the level of economic activity and the rate of economic progress. But to me the present contribution that theory makes to the analysis of these problems is not far short of trivial. Arguments between economists advocating one policy rather than another can usually be explained more significantly in terms of politics rather than economics. They develop into unedifying slanging matches in which each faction picks out those particular elements or that particular formulation of the problem which lead to the conclusion it favours.

I now turn to a rather different group of notions which we get from theory, but which do not imply theoretical propositions in the same way as those which I have just been discussing. For want of a better term, I call these common-sense maxims about the economic facts of life. Some of them are axioms of logical behaviour, others are general truths to keep in mind in trying to understand what goes on in the real world. Some in the first category are 'bygones are bygones', 'you cannot have your cake and eat it', 'in full employment, other things being equal, using more resources one way means using less in others', 'real costs are opportunity costs'. In the second category are such general truths as 'all transactions are two-sided; for every purchase there must be a sale', 'the balance of payments must balance', or slightly more subtle ones, 'changes in stocks are merely another way of expressing the relation between output and consumption', and the one made famous by Marshall, '*Natura non facit saltum*'.

I have always found that propositions of this kind play an important role in understanding what goes on in economic life and in arguments about economic policy. Take one or two examples. The notion that with resources fully employed you cannot have more of anything without giving up something else, played an extremely important part in all wartime economic planning. Arguing about the choice of one thing rather than another, explaining the cost of the choice in terms of other potential uses of the resources involved, ensuring the consistency of different choices, was in all its various aspects the day to day activity of many economists temporarily employed in the Civil Service in wartime. In this activity they were using little more than the simple maxims of consistent logical behaviour, arguing that two and two make four and never five. True, the working out of this logic in practice was most complex, for it involved taking account of time, for example in considering the relation between the build-up of munitions factories, munitions output, and the size of the armed forces, and it implied a knowledge of facts, for example about the ease with which resources could be moved from one use to another; but the general principles or maxims being used were of the most elementary kind.

Or take the maxim that 'bygones are bygones'. It was because economists used this notion that there was so frequently a difference between them and the accountants, civil servants, and business men about the advisability of Government encouragement for extensive investment in the cotton industry after the war. In considering the worthwhileness of investment in automatic looms for instance, the accountants usually insisted on the comparison being made on the basis of newly purchased automatic and non-automatic looms. The economists on the other hand argued that there were plenty of non-automatic looms already in existence, long since written off, but usable for many years ahead, and that the estimates of capital costs on non-automatic looms should, therefore, be taken very near zero.

Of the general truths which are important to keep in mind in trying to understand what goes on in the real world, I would pick out for special mention 'the double nature of economic transactions' and '*natura non facit saltum*'. The importance of these can be judged from the error into which people are led by ignoring them. How frequently do people discuss the potential threat to the rest of the world of increased exports from the European Common Market without paying attention to the implication that this would mean increased imports into Europe too. Or refer to the opportunities open to some British industries for expanding exports in the European Free Trade Area—when this was still a possibility—without realizing that this would mean increased imports into this country. How frequently do students talk about sales of gilt-edged, and appear rather nonplussed when asked who buys them from the sellers? Or do you remember the immediate post-war period, when the use of the word 'inflation' was still taboo, and Government propaganda—from the Ministry of Labour particularly—kept on plugging the idea that we were suffering from 'an overall manpower shortage'. If we could only increase the labour force all would be well! In the way the argument was presented to the public, the fact that this would also mean increased incomes, was completely overlooked. All simple errors, but how easy to fall into them, and how disastrous the consequences if policy is based on misapprehensions of this kind.

The Marshallian maxim that '*natura non facit saltum*' is ignored in the 'all or nothing' view which the public usually takes of economic affairs. Such slogans as 'Export or Die,' or popular assessment at the time of the Suez crisis of the economic importance of the Suez Canal, are based on a complete misunderstanding of the nature of economic activity. The public seemed to be under the impression that the closing of the Suez Canal would bring British industry to a standstill. In fact adjustments at the margin in fuel supplies and consumption, and in the use of alternative shipping routes, together with the availability of stocks, meant that the effect was quite small. It is indeed now very difficult to pick out the 'Suez crisis' period in the figures of production and trade. Here again the way in which economists look at these issues, in emphasizing that the possibilities of substitution at the margin are many and varied, and that issues of policy are really about marginal changes, is of great value in correcting public misunderstanding.

Lastly, in this section on theory, I turn to theoretical concepts which I think of as a special shorthand language which economists have invented for describing economic relations. Many of these play an important part in the theoretical models which I discussed earlier, but I am not here thinking of them in that context but as parts of an economic language unrelated to any particular theory. Such terms as elasticity, mobility, competition, substitution, short-period, long-period, net advantages, equilibrium, marginal and so on. These are used in applied economics and discussions of policy mainly as a shorthand language. And in my view they normally contribute no more than this. They do, of course, make it possible for economists to talk to each other in a jargon of their own, which excludes outsiders from the conversation, and in some circumstances this gives the profession a certain prestige. This is perhaps in itself a fairly harmless form of vanity, but the position is more pathetic when economists by using the language merely deceive themselves. For the use of the language may give the illusion of great understanding, whereas in fact it often merely conceals ignorance in a mass of esoteric jargon. And it is so easy and dangerous to mistake description and classification of situations in a special economic language as answers to problems.

So much for theory. I now turn to the other kinds of knowledge used in applied economics. These are mainly factual, and are of two kinds, statistical and institutional. The information about institutions may be merely descriptive or may also attempt to give some understanding of how the institutions work. I have less to say about the application of this kind of knowledge than about theory, for I find it even more difficult with this to distinguish between the illusion of understanding and real understanding.

Take statistical information first. There is, of course, a passion for statistical information in relation to any and every issue of economic policy. Indeed the normal first reaction of any economist today considering a real problem for the first time is to complain of the inadequacy of the statistics available. And yet how often do we really honestly ask ourselves what we get out of the figures? True, they give us a comforting feeling that we know a great deal, and they enable us to take ourselves or others on an impressive statistical tour of some part of the economy—what we usually call, 'an appreciation of the situation'. Once this kind of activity reaches a certain level, it grows and spreads by its own power. For Government Departments, business, and organs of publicity and education feel the need to have economists who can write appreciations for them and argue about and interpret those made by others. But where essentially does it get us? Are we any better off than the commentators on politics and foreign affairs who can give us exquisitely neat and tidy analyses of the current political and international scene, but can tell us precious little about what we should expect to happen next, or what is the best way of dealing with the situation?

Take the analysis of the present economic situation. We can write a detailed statistical appreciation quite readily, drawing attention to all the important elements in the situation. This is no doubt of value in telling us where we are, but it gives us precious little clue to what is going to happen next. I am not wanting to decry the importance of the knowledge that we get from statistical appreciations of this kind, but merely to suggest that we should be careful not to kid ourselves, let alone others, that by the mere accumulation of such appreciations comes greater understanding.

At least, however, appreciations of the general economic situation are directed to helping Government or business in decisions which they have to take. What about statistical appreciations which have no obvious pointed direction of this kind? I well remember asking myself this question after giving a lecture a few years ago, called 'The Economist looks at the Coal Industry', to the Conference of Colliery Managers. I had warned the man who asked me to do this that I knew nothing about the coal industry, and he said, 'fine, just what we want. A fresh mind, with no preconceived prejudices'. In the end I succumbed and gave what I suppose is a fairly typical paper in applied economics. Apart from deploying the usual arguments for changing pricing policy for coal—arguments which used the elementary theoretical notions I discussed earlier—my main themes were statistical, although these were filled out with quite a lot of patter. Some of the themes I dealt with were: the importance of coal in the economy as a whole, the effect on the balance of payments of the loss of coal exports, the main trends in coal consumption, fuel costs as a proportion of total costs in industry, the importance of fuel in consumers' expenditure, movements in coal prices compared with other prices, and so on. I do not believe I am being arrogant when I say that the lecture went down well, and there were many comments afterwards on how enlightening my talk had been. And yet as I drove back to Manchester I found it difficult to answer the question—in what way enlightening? I had put to my audience a number of facts about the coal industry of which they were obviously previously ignorant, and no doubt this enabled them to see the problems of the coal industry in a broader perspective. I had also attempted to correct some mistaken popular notions, for example about the rise in coal prices compared with other prices since before the war. But was there anything beyond this?

Indeed a good deal of the use of economic statistics by the applied economist in public discussion seems to be of this negative kind—correcting mistaken notions of facts which have for some mysterious reason bitten deep into the public mind. When politicians and others go around arguing that we are in trouble because we do not pay our way overseas, one can quote the balance of payments figures of the last ten years. One can attempt to correct mistaken popular notions about relative rates of economic progress in the U.S. and U.K. since the war, by quoting some of the statistical evidence. In the Suez situation which I have referred to earlier, it would have been very valuable if one could have got across to the public some of the statistics about the fuel and shipping situation. And as no doubt misconceptions of fact will continue in the future, as in the past, the economist who is sufficiently interested and familiar with the statistics will continue to find a useful application for his knowledge in telling others what nonsense they are talking.

Economic statistics sometimes play a role in economic policy which is different from that which I have been discussing, and although not directly related to my central theme, it may nevertheless be worth mentioning. I have in mind those statistics which assume great importance in the public mind and may influence the public a great deal in the action they themselves take or the action they demand from the Government. Take two sets of statistics of this kind which have significance, the figures of gold and foreign exchange reserves and the unemployment percentage. The evidence before the Bank Rate Tribunal showed what an important influence figures of the reserves had on the views of people in the City of London about the underlying trading position of this country, and therefore on their actions and decisions. Indeed one is sometimes tempted to take the view that there might be fewer speculative balance of payments crises if the gold and foreign exchange reserve figures were not published. Again the unemployment percentage appears to have a most powerful influence in politics. Great significance is apparently attached to whether or not it is below the magical 3 per cent. The fact that the figure is quite differently calculated and has quite a different significance from before 1948 and pre-war, is completely overlooked. Indeed the public reaction to statistics of this kind is something which any astute economic politician is bound to take into account. But I doubt whether he would get much help on this from the applied economist. Here is a field for the economic psychologist or sociologist.

The other kind of factual knowledge is mainly about institutions. A good deal of applied economics, in banking, finance, industrial relations, and industrial organization is descriptive of institutions and how they work. Let me take the organization of industry as an example, for I am more familiar with this than with the others. In what is written about industrial organization, there is usually a little potted history and technology, some statistics about the size of firms, production, and outlets, and then usually some attempt to discuss whether the industry is efficient or not. This may involve some attempt to analyse the extent of competition, restrictive practices, and the use of monopoly power. This is, for example, a fairly standard pattern in that extensive, but much delayed, symposium, edited by Duncan Burn, on the Structure of British Industry.

Here again my reaction is that the nature and usefulness of the understanding one gets is of a most elusive kind. Usually if the job is well done one gets the impression, I think quite rightly, that economic organizations are most complex in their variety and ways of working, that it is dangerous to think in terms of simple generalizations, and that sweeping proposals for reforming an industry or organization are dangerously deceptive. If this view is correct it means, again, that the usefulness of this kind of understanding is largely of a negative kind. For it merely enables one to talk with some apparent authority to people who are obviously ignorant about the affairs of the industry, and to comment on and criticize proposals for public intervention of one kind or another.

The one thing which is usually missing from such analysis and understanding is any reference to problems of internal organization, administration, and decision taking, and therefore any clue to what makes some firms and institutions efficient and successful and others inefficient and failures. Certainly this is true of my own contribution on the 'Aircraft Industry' to the Burn symposium. I spent quite a lot of time on this and can now pose, compared with those who know nothing, as something of an 'expert' on the economics of the industry. But on such important questions as 'what has made Rolls-Royce such an efficient and successful firm?', I am as perplexed as ever.

I now come to my third section. How does one teach applied economics? This is the most difficult question of all and the difficulty for me at least is partly due to the fact that I have not been successful in answering clearly and precisely the prior questions about what it is we are applying.

If you agree with my general argument that what we are applying in theory are the elementary propositions and common-sense maxims, it might seem to follow that it should be fairly easy to teach students of economics to discuss real economic issues critically and intelligently. But this is not so, in my experience at least. Students seem to be able to master theory and even to deal with notional applications of theory, but when it comes to discussing some real problem on which they have not been specifically instructed, they seem to be influenced more by notions they derive from TV or the *Daily Mail* than anything they have learnt in Economics.

Clearly one of the most difficult things to achieve is to get the student in a position in which the elementary propositions and maxims are part of his normal processes of thinking, and are not kept in a separate compartment labelled 'economic theory'. One way of doing this is to take him through examples of the theory in as many different contexts as possible. The laws of supply and demand in relation to commodities, factors of production, location of industry, international trade, etc. It also seems to be true that the elementary propositions will not be absorbed if the theoretical instruction is merely kept to the elementary level. It is only when the student has been through a theoretical drill well beyond the elementary level that the elementary notions really sink in deep.

But it is, of course, not enough to learn the theory as theory. In some way the bridge to application to real economic situations must be crossed. It is here I think that at present we make the biggest mistakes. First because there is a tendency to show how the theory can be applied to get solutions, whereas, to my mind at least, we ought to concentrate on showing how the theory illumines the nature of the problems. Secondly because it is the big issues, the control of inflation, devaluation, agricultural policy, nationalization, that are given to him as the important issues of application. He is expected, judging by examination papers, to provide solutions to the most difficult current economic problems. At the end of three years, if no earlier, the student is encouraged to see himself as an exceptional Chancellor of the Exchequer, who could put everything right if he were only given half a chance.

I think this is disastrous, and that we must in some way try to teach him applications of a less ambitious kind. The trouble is that these are not very easy to find, for most of the literature is about the great issues of public policy. But the two examples I gave earlier, of the operation of the elementary propositions in supply and demand theory, are the kind of examples I have in mind.

The other thing we should do is to make the student suspicious of slogans in economic policy, and to make him understand why he should be suspicious. But it is my impression that, far from doing this, many teachers of economics, in fact, feed their students with slogans: 'Invest more', 'Increase the Gold and Foreign Exchange Reserves', 'Double the standard of living in twenty-five years'. I do not want to argue whether or not these are good slogans for policy makers. My main point is that this is not the way to turn the young student into an applied economist, capable of thinking for himself.

The other puzzling question is how to get across to the student some of the knowledge both statistical and institutional that I talked about in my second section. I have tried to get students to learn to read economic statistics, and to write statistical appreciations about some aspect of the working of the economic system, but I have not been very successful. I am forced to the conclusion that this aspect of understanding, for what it is worth, is acquired only by long painful apprenticeship, and with some, as is only too patently obvious, it is never acquired at all.

In trying to convey an understanding of how institutions work, one is perpetually up against the difficulty of discussing problems of economic administration with students who have never had experience of an administrative problem in their lives. And to those without such experience such problems are usually either completely baffling, or obviously the result of stupid inefficiency. If the student can be induced to get a job in a firm or

Government Department for a few weeks, not to earn a lot of money, but to see something of the working of an institution in practice, it is much easier to discuss such problems with him afterwards.

Before I conclude I feel I must apologize for not having answered more adequately the questions I posed at the beginning of this paper, although I hope I have given my readers something to argue about. I do really feel that in twenty pages of rather discursive argument I have got little beyond what Keynes stated so succinctly in the first two sentences of his introduction to the Cambridge Economic Handbooks nearly forty years ago— 'The Theory of Economics does not furnish a body of settled conclusions immediately applicable to policy. It is a method rather than a doctrine, an apparatus of mind, a technique of thinking, which helps its possessor to draw correct conclusions.'

Postscript

In pondering over the discussion of the above paper, as read at Southampton, I think the following points worth noting. These are not meant as a summary of the discussion, but merely some of the points I would try to cover if I were rewriting the paper.

1. My paper was not meant to be 'an attack on theory'. I was merely trying to answer the question 'what part of *present* theory is enlightening and usable?' If anything, my remarks about the use of economic statistics and descriptive economics were more damning than my comments on theory. For I argued that there are useful elements in theory, but that I saw little point, except of a negative kind, either for understanding or for policy, in most statistical and empirical descriptions of the economy.

2. By what criteria does one decide whether theoretical propositions are 'elementary and simple'?

Is it true that theory which is complex and difficult to use for the initiating generation, is simple and useful for the next generation (e.g. by analogy with physics theories taught today to schoolboys of sixteen were complex and difficult even for eminent physicists thirty years ago)? That theory which is apparently complicated and unusable today, will appear simple and useful to economists in thirty years' time? If this is so, then my argument that only elementary and simple theory is useful, would, to some extent at least, be tautological.

I do not think that this is so. It is not that theory becomes useful as time passes, because the theory, originally thought complex, progressively appears simpler as it is absorbed into general thinking. My argument is that, whether or not this happens, it is the simpler aspects of the theory which are usable. I admit that it is difficult to disentangle complexity and simplicity from strangeness and familiarity. But in price and value theory, for example, although much of what was strange and difficult twenty-five years ago is familiar and easy today (i.e. what was advanced economic theory then now appears in first-year courses), it is still the simple and elementary aspect of what continues to be a complex theory which is usable. (Price theory would appear strange and complex to an African brought up in a primitive economy with a traditional exchange system, but when he had got over the strangeness it would still only be the elementary aspects of the theory, i.e. the general laws of supply and demand, that he would find useful).

3. What do I mean by 'theory'? This is obviously not easy to answer. Certainly not just any general idea or proposition with which one approaches the study of reality. In this sense theory is always involved, if merely by implication. Even in an apparently theory-free description of the facts, some general propositions are implied in deciding which facts are worth including in the description.

By theory, I had in mind the formulation of a model in which the relation between the elements being considered was expressed, or could be expressed, mathematically. This is what to my mind 'economic theory' has essentially been in the past, and continues to be today.

In attempting to understand reality, one can use models of this kind. The models may or may not be formulated in such a way that they are in principle testable. Even when in principle they are testable, there remain the complications, statistical and other, in relating them to complex reality.

The alternative approach to understanding is what in the discussion at Southampton I called 'the historian's technique'. I do not mean by this trying to find 'laws of trends in history', but the approach to reality which tries to understand what is going on by 'soaking oneself in the facts of the situation'. In this approach one has no precise model clearly formulated in advance which one is testing against reality. One may, however, have certain general considerations in mind of what are important elements in the situation, and one approaches the facts within this kind of vague framework.

4. Is further progress in economic understanding, which will be usable for policy, more likely to be made by the 'theorist's' or the 'historian's' technique? There is clearly no *a priori* logical basis on which one can argue that one will always be more fruitful than the other. The proof of the pudding must be in the eating. But there are certain characteristics of the two which are of some importance:

(a) The theoretical approach usually implies the specification in advance of the model which is to be tested. The danger here is that in the testing process facts that do not fit into the framework of the model will be ignored. It is difficult both to specify logically the model in advance and to be flexible in using it in studying reality. The ideal of 'a closed model and an open mind' is difficult to achieve.

In any case a theoretical model cannot be based on mere introspection and thought about logical relations. It must have some relation to reality, and the appropriate elements of reality to assume can only be selected by some process other than that of model building.

It does not seem to be possible at present, sensibly, to formulate some of the problems in which we are interested in the form of theoretical models of this kind. If, therefore, we insist on this method as *the correct* way of conducting research, we either produce rather silly models or say that we cannot deal with these problems.

(b) The 'historian's technique' is open to the danger that since we do not clearly specify in advance what we are looking for, we run the danger of merely collecting facts for facts' sake, and may end up without learning anything very useful. Even where we have some general propositions vaguely in mind, the element of personal judgement in assessing the facts in relation to these propositions may be very large, and therefore two people approaching the same facts may come to quite different conclusions. This indeed very frequently happens among historians, each historian giving his own interpretation.

5. There are many problems which to my mind cannot be sensibly investigated at present with the theoretical approach. Take two which were discussed at Southampton:

(a) First the relative importance of 'security' and 'competition' as elements in economic progress. No doubt some ingenious theorist could formulate a model which included these as dynamic variables and could produce a solution giving the optimum situation. But I find it difficult to envisage such a model being useful either as a research tool or for policy.

If one wants to investigate this problem one must, at present at least, proceed by the 'historian's' method. That is, try to examine, in relation to a particular industry or group of industries, what the relation between the factors appears to be. In doing this one might try to proceed by the comparative method, but this would involve much judgement. For it is unlikely that one would find a series of pairs of industries apparently similar in all respects, except that the firms in one experienced great uncertainty about the future because of fierce competition, and in the other enjoyed great security because of monopolistic structure or restrictive agreements. One must be prepared to conclude that in some situations 'security' and in others 'competition' appears more conducive to progress, without being able to specify in general terms what the differences in the two sets of situations are.* The results for policy making would then be, as

* Indeed one must also be prepared to emerge with a conclusion that neither factor, competition, nor security, had much bearing on economic progress.

Applied Economics: The Application of What?

they frequently are, that 'each situation must be judged on its merits'. Such a conclusion would not be very helpful, but I cannot envisage that proceeding by trying to test more formal models would at present get us any further. This does not mean, however, that those who think they can get further by this route should be discouraged from trying.

(b) Second, one problem I mentioned in my paper, 'What makes Rolls-Royce an efficient aero-engine firm?' Here, to my mind, we do not have any of the elements for formulating even a tentative model in advance. One may have vague ideas, e.g. that it has all depended on the personality of Lord Hives (the great man theory of business efficiency?), but these are so uncertain that even in principle I do not see how one could specify in advance what general propositions one was going to try to test. My view would be that if one was attempting to answer a question of this kind, one would have to go and work inside Rolls-Royce for some time, starting with an open (almost blank!) mind, only formulating views as one progressively soaked in more and more evidence about what made the firm tick.

6. Given the views expressed in (4) and (5), there cannot, to my mind, be any logical guiding principles on which to allocate resources between different methods of research. This must inevitably be a matter for personal interest, capacity, and judgement (sometimes called prejudice).

15

Some Notes on 'Philosophy of History' and the Problems of Human Society

D. M. MACKINNON

1. *Introduction*

WHILE it is perfectly true that most political theorists have largely neglected the issues raised by the involvement of individual human societies in an international community, the same is certainly not true of some of those who have essayed what may be comprehended under the term 'philosophy of history'. I say what may be comprehended under that term, for between the variety of enterprises which the expression may indicate, there seems at first sight little in common. Moreover there are not lacking today many voices of unquestionable academic authority, who would query the status of most, if not all, the enterprises that might be thought to shelter under the umbrella-term 'philosophy of history'. They exist in a kind of twilight realm, an intellectual *demi-monde*, lacking the respectability properly claimed, on the one hand, for first-hand historical research, and on the other, for attempted characterization of the logical principles involved in such researches. A 'philosophy of history' is, it is argued, as bad a guide to the determination of what actually happened in human societies as an alleged *a priori* evident metaphysic to the course of nature, and the structure of the universe around us. True, the development of the natural sciences is in debt to the wayward styles of cosmological speculation; true, the historian may be in debt to Karl Marx for compelling a revised estimation of the relation of constitutional to social and economic history. But a 'philosophy of history' is something from which men must *extract* insights, whether those insights be individual historical reappraisals, or corrected perceptions concerning the proper methods of historical inquiry.

Yet the student of the history of ideas cannot let the matter rest there; especially if that student counts nothing human alien from him. For 'philosophy of history' remains the name of a variety of enterprises to which men have found themselves impelled; one can trace the presence of something inviting the use of the term in Paul and Augustine, in Bossuet, in Adam Smith and Burke, in Kant, in Hegel and Marx, in Acton; one could continue indefinitely. It is not that in these very different writers, the use of the term indicates some isolable common concern, as one way discern in an exactly identical symptom in a group of sufferers from the same disease. A group of children suffering from measles may have, at the same time, rashes of a more or less identical quality. Rather the kinship is more akin to what Aristotle in the last chapter of his *Categories* discerned to obtain between different forms of having. A man may have five shillings in his pocket, an unread copy of Russell's *Principles of Mathematics* on his shelves, expectations from his aunt; he may have a devoted wife, a son in a comprehensive school, an uncle serving a sentence in jail for fraud. What is here in common between the various forms of having? Are some forms more nuclear, nearer to the heart of the matter of having (as Aristotle believed *substance* to be nearest the heart of the matter of being) than others? Is there a frontier traceable between the literal and the metaphorical uses? The inquiry is tantalizing; but I mention it only by way of illustration. When we say that in a man's thought we can discern something we call 'a philosophy of history', we do not mean to suggest that it always fulfils the same function, that he invokes its resources to make precisely the same move in some precisely comparable game of intellectual chess. It is even true, where a writer whose total *oeuvre* was as massive and as complex as Hegel's, that the term indicates a number of very different concerns. Those who knowing only the Hegel of the *Encyclopaedia*, come to the young man, slowly achieving, out of a mass of discrete and discursive reflections, sometimes nearly *existentialist* in temper, the structure of his *Phenomenology of Spirit*, are brought up sharp by the contrast between the tentativeness of his *Jugendschriften* and the self-confident architectonic construction of his later years. Yet it could be argued that for both the young and the mature Hegel, something called 'philosophy of history' lay near the heart of his perplexity. One can well ask how far the term indicates, in both periods, an identical enterprise.

2. *Plato and Thucydides*

In a recent paper Mr. Martin Wight of the London School of Economics emphasized that among the ancients, it was Thucydides, who before all others, revealed an obtrusive preoccupation with the problems raised for men by the extent to which that society was involved in an international community. Now even if we discount Cornford's attempt to find in Thucydides's *History of the Peloponnesian War* 'an Aeschylean drama in a shaggy prose', attempts are not wanting to extract from Thucydides a 'philosophy of history'. These may take the shape of something which hardly goes beyond the level of finding in him sustained and highly generalized argument for his own political preferences; or they may approach (even asymptotically) Cornford's oft-quoted verdict. Certainly the present writer (who is no classical scholar) would not dare enter these lists himself, were it not that the effort to understand Plato's *Republic* had first introduced him to David Grene's excellent book *Man in his Pride* and then compelled him to return to the study of Greek history, and especially to the figure of Pericles.

The *Republic*, Mr. Wight argued in the paper just mentioned, helped to fasten upon us the tradition that tied political theory to the exploration of the grounds and limits of the claims upon us of the laws of the society under which we live. Certainly he would have to make a curious concession in Bk. 4 to the realities of power politics; but the self-contained character of the *Kallipolis* so stressed by Plato, helped to shut out a conscious awareness of the context in which any society must subsist, from marring the order of his exposition. But the question, however, remains: What in the last resort is the *Republic* about? And here I want to hazard a suggestion that may seem absurd. It is, in part, at least an attempt to answer the question raised in the minds of the young men, Glaucon and Adeimantus, by recollection of the opposed ways of life of Pericles and of Socrates. That the picture of the *perfectly just man* of Bk. 2 is suggested by Socrates is commonly admitted; it seems to me arguable (especially if one attends, for instance, to the last speech which Thucydides puts in Pericles's mouth in Bk. 2 of his *History*) that the *perfectly unjust man* is to be seen in Pericles. Note that the perfectly unjust man is not the successful hypocrite; his *seeming* just should be understood with all the force with which the *Republic* invests the notion of *seeming*, in mind. That *which seems* is between *what is* and *what is not*. The perfectly unjust man does things which are just; the list of his achievements is evident for all to see, including the welfare he has secured, the public works he has sponsored and made possible, even the temples and sacrifices he has endowed, and these are no trifling things. But they have been done at a cost, even as the Parthenon itself had been built out of money, not required after the peace of Callias, for the purposes of the Delian Confederacy. True, Athens had risked her land and her temples in the common cause, and there was undertaking that her losses should be made good at the expense of the common foe. But the years 449–8 saw a profound crisis in the relations of Athens and her allies, a crisis reflected in the internal politics of Athens. One can argue (the facts can be marshalled) that those allies, even after the reign of the demagogues, and the Sicilian débacle itself, remained loyal to that State whose protecting power had nourished the growth of popular government within them. But while the ambivalence of the imperial idea is there, its kinship with the horrid reality of *tyrannis* cannot all the time be evaded.

If there is anything in the interesting arguments of Dr. Victor Ehrenberg's study of *Sophocles and Pericles*, the *Oedipus Rex* and the *Antigone* reflect, in differing ways, preoccupation with this same problem raised by the person who seemed to incarnate the Athenian idea, in its ruthless energy as well as in its delicacy and richness of achievement. But Plato eschewed the styles of tragedy as decisively as he eschewed the styles of Pindar, whose ninth Pythian Ode comes nearer than the whole *Republic* to embodying such a comment on the Periclean age as a Burke might have passed. He sought to offer men less a comment to illuminate their minds than a method of proof which would finally banish from their minds the seeds of a final moral scepticism. He would bring men to a place from which they could see the superiority of the way of Socrates, and recognize for what in its inmost core it must be, the way of the tyrant-man. His temper is not conservative. He is attracted by elements in the manners of an older Greece, even as he admits a debt to Spartan institutions in his educational programme. But he acknowledges a deeper debt to the intellectualist temper of his native Athens, and with Pericles he would disdain the human wastefulness of the Spartan system, the exaltation of toughness and endurance into a way of life which is wrongly judged its own justification. He is altogether radical; and his radicalism issues from his sense that the deepest moral scepticism (for the scepticism of Glaucon and Adeimantus is more profound than the self-confident 'placing' of morality in the scheme of things, set on the lips of Thrasymachus) is born of a sense of the decisiveness of amoral power in the order of the world. What counts among men, what entitles men to due measure of praise, is built upon a foundation of superiority on the level of force, of power, of resources. Of course such superiority does not by itself achieve a due energy and creative enterprise in their use. A man may lack the will to convert his advantages to proper use; but such advantages are the precondition of achievement.

Do we have here what may be called 'philosophy of history'? One might say that we have here a classical example of another sort of response to that kind of awareness concerning the human situation which men have tried to express, and to come to terms with, by means of a 'philosophy of history'. I mentioned a few moments ago Ehrenberg's thesis which he developed with the aid of some very interesting detailed fact (relating e.g. to Pericles's and Sophocles's colleagueship in the strategia) concerning the possible political impact of the *Antigone* and the *Oedipus Rex*. Sophoclean tragedy was a medium, an art-form in which Hegel (if I may leap the centuries) discerned the very substance of his 'philosophy of history'. Antigone, Creon, Oedipus, all in their several ways, are seen by Hegel, estranged from the proper order of human life by their achievement, not by their weakness. Ehrenberg's assessment of the *Antigone* is, of course, very different from that of Hegel: this because he brings together the figures of Creon and of Oedipus as both alike embodying Sophocles's profound comment on the tragic corruptibility of kingship. (Hegel, notoriously, presents Creon and Antigone as dialectically opposed thesis and antithesis in the deployment of the relations of society and self-conscious individual). Pericles is *prostates*, and by his *prostasia* with all it made possible for the 'démocratie dirigée' of his city, he became infected with the deadly sin of *hybris*, the *hybris* of Creon, the profounder guilt of Oedipus. And as, if Ehrenberg is right, Sophocles wished to convey to the audience at the Great Dionysia, the virus of this guilt is at work in the blood-stream of the men and women of the city wherein he is *prostates*.

By tragic drama, men can learn the perils that must beset them; they can recognize, come to terms with the ambivalence of their situation. It offers a method of *exetasis*, of examination, whereby insight may be deepened into the contradiction of the human situation. But Plato rejected it, maybe because it encouraged in men the habit of a false pity, because he feared its implication might be a too facile *Tout comprendre, c'est tout pardonner*. To speak of Pericles as a tragic figure suggests that one is already on the road to extenuating his guilt. The spectator of a series of tragic dramas is more inclined to be a *philotheamon*, one who passes from spectacle to spectacle, enlarging his sympathy no doubt, but ill-served if he supposes that by so doing he deepens his insight. Depth of insight can only be achieved by that concentration upon the *unum necessarium* that is the way of philosophy, whereby a man turns his back upon history, and upon the cultivation of a tragic sense of life, and indeed upon what one might be tempted to call 'philosophy of history', namely the achievement of that latter sense by a certain manner of representing, of selecting and then portraying, the substance of historical occasions and individuals. But 'philosophy of history' is neither philosophy nor history. Certainly reflection on historical actualities, the passion of Socrates, the *prostasia* of Pericles, can furnish men with the *point de départ* of their philosophizing. But the philosopher is not a commentator, but a judge; his aim is not to enlarge pity, but to secure truth.

3. *Providence and Tragedy*

History: tragedy: philosophy—to understand what we call 'philosophy of history' we must (and here our debt is great to Hegel, however much we must reject, e.g. his interpretation of individual tragedies, such as the *Antigone*), see tragedy and philosophy (as Plato understood it) as alternative responses to the same sort of perplexity, and recognize in 'philosophy of history' something that *may* have in it something which can only be described as the presentation of the tragic *sub specie philosophiae*. I say: *may*—for do we discern in Sophocles anything remotely akin to the notion of *providence*, that notion which, discernible, e.g. in Burke and Adam Smith as well as in Kant, represents a kind of distillation, an aetiolated transcript of a whole constellation of Hebraic–Christian ideas? Few questions in the field of our exploring are more fascinating than that of the relation of the Christian *kathos gegraptai*, when it is said that 'the Son of Man goes *kathos gegraptai peri autou*, but woe to that man by whom the Son of Man is betrayed' and the tragic element in human life, the intractable backlash of Sophoclean thought (In the Passion-narrative of the Fourth Gospel the theme of the role of the state in the crucifixion of Jesus is made the occasion of a supreme exercise of the Evangelist's irony; but it is not tragic irony. Pilate has no power against his prisoner except in terms of Caesar's mandate; he plays the role assigned to him because he is governor, and it is as governor (who bears not the sword in vain) that his prisoner confronts him. But of course he plays another role still for which he has been cast by the One whose emissary the Son is in his faithful witness to the Truth. And in the end the reader comes to rest in the reality of Providence, a providence conceived not, however, abstractly, but made concrete in the history now drawing to its close.)

Providence: I have mentioned an aspect of the Johannine presentation of this theme. Recognition of providence, detached from the minute theological, even Christological, particularization it receives in the Fourth Gospel is the general category whereby the tragic element in human life has been made bearable. Theodicy, the justification of the ways of God to man, has always, since Paul, since the Gospel of Mark, lain very near the heart of Christian theology. The scandal of a crucified Messiah—that was the problem, and the solution. But the 'great disturbance' in the ways of human thinking, that this scandal provoked, bore fruit in the enormously varied 'Idea of Providence'. One could say of Hegel's philosophy of history (and his debt, e.g. to Adam Smith, whose theory of the 'harmony of interests' he saw as a peculiarly pointed illustration of his doctrine of the 'Cunning of the Idea') that it represented an attempted synthesis of the notion of tragedy and of providence. 'Pas un panlogisme; plutôt un pantragisme'. So a French writer on Hegel. What then makes it possible for a modern critic of Hegel like Sir Isaiah Berlin to omit altogether the emphasis on tragedy from his classification of Hegel as an historical determinist? Simply that in his doctrine the notion of the tragic (so vividly present in his explanation of the idea of alienation, in his criticism of Stoic and Kantian ethics) becomes gradually subdued by his sense of the providential ordering of human history, of the mysterious tactic that disposes all things for good.

It was part of the diffused inheritance of Christianity that it delivered men out of a world of tragedy, a world in which there was always a backlash, an intractable element, resistant to penetration by the rational form a creator would impose upon it, into a world of providence wherein all things work together for good to men that love God, a world in which *theologia crucis* was the veritable *theologia gloriae*. Hegel wrote at a time when the authority of the Christian story was diminished and brought low, when indeed also the Lisbon earthquake of 1755 and Voltaire's *Candide* had impressed anew upon men's minds the alien quality of their *natural* environment, and even Kant, for whom the reality of autonomy had been the very foundation of his ethical doctrine, had found himself driven to mitigate the sharpness of his separation of the *Realm of Ends* from the *Realm of Nature* by allowing validity to belief in One in whom their

Some Notes on 'Philosophy of History' and the Problems of Human Society

reconciliation was assured, even if the control of that reconciliation was inconceivable. If the notion of *providence* (which has its role in Burke's political theory, whether as the ground on which he bids men accept the American or the Irish reality, or urges Jacobins to refrain from identifying their partial insights with the very laws and orders of creation) is to be rehabilitated, it is only a *providence* that is seen as operative dialectically, that is tragically, whether the tragedy be discerned in the triadic structure of the *Oresteia* or in the movement of Christ's ministry from life to death to Resurrection, from Galilee to Jerusalem to Galilee. Marx blamed Hegel for supposing we must seek to *understand* the world; whereas our task was to change it. But what was the nature of the *understanding* Hegel was blamed for seeking? Surely it was that reconciliation with one's fellows, with one's environment, natural and social alike, which tragic *katharsis* (not necessarily conceived as by Aristotle) may be thought by some to bring. One entered into a kind of communion with the order of the world, seeing it as a macrocosmical expression of the law of one's own spirit. And in this communion was acceptance, reconciliation: a false acceptance, for Marx, a reconciliation which was an escape.

It is because in Hegel, and in Marx, his critic, the impact of the tragical upon the providential is set out and criticized, that their thinking has a peculiarly radical quality. What is the role of the tragic, of that element in human things which drove Plato to *philosophize*, to set out the system of that education which alone could equip men to bear the burden of the debased life of the cave, in a providential scheme? Of course, in the Christian ages, men had recognized the problem, dealing with it now one way, now another. One can see perhaps in those sections of Paul's writings in which (as in 2 Corinthians) he turns in upon himself, to offer an account of the *paideia* of an apostle analogous to the Platonic exposition of the *paideia* of the philosophers, the movement towards the transcendent that is always one expression of the admission of the unbearable.

Moreover, in writers on ethics and politics like Butler and Burke, where the sense of the mysterious ways of providence, the unknown bringing sense out of nonsense, enters as a kind of extra dimension into their theory, the intrusion is always woven tightly together with a renewed expression of the medieval tradition of 'natural law'. There are constants in human life, whether they be found in the complexity of our actual nature, or in the strange particularities of human history, where none the less, a properly disciplined eye can discern moral universals. The sense of the *abyss* which yawns as soon as one insists on the reality of the tragic is absent. To recognize the ultimately intractable can breed in men a kind of pity that inhibits action, or can compel them along the kind of road that Plato set out in the *Republic* (the flight from history, if you like, to which Christian eschatology ever, somehow, barred the way of Christian practice), or else into that identification of tragic sense with ultimate understanding in which Marx saw Hegel's perpetuation of the worst fault of religion.

4. *The appeal of Marxism*

It is the temptation of the philosophical libertarian to forget the dynamic energies that are released by the sense of providence, of a providential order of which one is oneself the agent. It ought to be sufficient to guard against falling into such temptation to remember the example of the sixteenth- and seventeenth-century Calvinists. To understand the continued appeal of Marxism (in spite of Hungary) one has to see that in Marx–Leninism men are urged to see that all they must do in themselves in order, in a style truly Promethean, to be themselves the architects, the Providence of a new world, is to purge out of their hearts lingering religious or quasi-religious attachments to the significance of the tragic. Their experience will be tragic, will have in itself in the burden it lays upon them, all that quality of the sheerly unbearable to which the tragic form bore witness; they will have to do things (e.g. in the execution of a five-year-plan) that seem to estrange them from their world; they will have even (if the Party demands it of them in a purge) to estrange themselves from themselves, to drive a wedge between their past and their present.

This will be their lot when men in the Party have made their own the truth that they (in that self-conscious direction of the human world which the Party makes possible) are the only Providence: the only Providence in relation to which men achieve and practise freedom.

Of course this kind of approach, this highly sophisticated perception of what it is that gives to Marx–Leninism its strength today as distinct from in the thirties, takes for granted the sort of knowledge of the history of political ideas a modern Western intellectual can command. The appeal of its Promethean promise to less sophisticated societies is obvious enough. There emphasis must fall on its *erfolgsethiker*, utilitarian readiness, to be judged by the visible, tangible fruits of its adoption as a social policy. Its apologetic will be Benthamite, vindicating the cost of industrial revolution by its fruits, more evident vindication of that price than can be overturned by criticism in terms of some abstract theory of 'human rights'. But the style remains humanist; men commend not a God that will surely fail them, but themselves as heirs, and indeed as effective, if not infallible heirs, of the promises of the religions of past ages.

All these remarks are fragmentary and ill-ordered; they are no more than a scratching of the surface, seeking to suggest (at best they do no more) the presence in some facets of political and social thinking of unacknowledged preoccupation with those mysterious depths of human life that poets and seers have tried to capture. We are all of us somehow compelled to come to terms with this 'problem of evil'; but sometimes our thinking and our action on other topics and in other places lacks clarity and coherence from our failure to disentangle the inevitable pressure of this problem on our unconscious imaginings and feelings.

16

Law-Courts and Dreams

ELIZABETH SEWELL

Poetic figures, which are figures of speech and figures of thought, are not fortuitous. Equally, a metaphor or simile or even a simple collocation of two things in a poem is not something finished and completed, like the answer to a sum or the conclusion of a train of logical reasoning. Such things, in poetry, are perhaps more like notations of a particular point reached in an active process of speculation, and are an invitation to speculate further. I want here to take one such collocation, simile, metaphor: law-courts and dreams, which occur together fairly often in literature as we shall remind ourselves in a moment. Poets set them side by side, in collocation; compare them, in simile—a dream is like a law-court, a law-court like a dream; or fuse them wholly in metaphor—law-court *is* dream, dream law-court. Thus public and private mind, forms juridical and forms psychological, are suggested as mutual interpretations of one another, by poets who in their turn have many times been called dreamers and who, one of their number claimed, are the unacknowledged legislators of the world.

The law, dreaming, and poetry—it is Shakespeare, and the Snark, who can start us off.

One of the most ironic and beautiful of Shakespeare's sonnets begins, 'Farewell, thou art too dear for my possessing.' It opens, as *King Lear* does, with an ambiguity or pun, a single word which turns like a hinge on a door between two separate mansions of our thought which yet, word and poet maintain, may communicate the one with the other. That turning 'dear' suggests that the laws of love and laws of economics may be connected. (Shakespeare would not be alone among the poets in suggesting this; there are hints of it in Novalis's *Fragmente*, and Shaw sets it down explicitly while claiming it as one of the great Wagnerian insights.) We shall not follow here this particular poetic sally between private and public affairs, any more than we can pursue the equally fascinating one in this sonnet's last line where the phrase 'In sleep a king' brings dream and monarchy into a connection which, once made, rings numerous bells through later literature, from *King Henry IV, Part II* and *The Duchess of Malfi* and Calderon's *La Vida es Sueño* through Grillparzer to Christopher Fry. That we dream up, as Americans say, our kings may well be true in some sense, with that corporate imagination of which little is known yet and which goes to the shaping or misshaping of all our forms of public life and government. For the present, however, we shall have enough with the connection between the law and the dream which the middle of this sonnet proposes.

It is a splendid farrago of legal terms. Estimates, patents, charters, bonds, misprision, judgements, follow hard upon one another. Then, 'Thus' says Shakespeare as if firmly to hook his legalisms home, 'Thus have I had thee as a dream doth flatter.' This is not the only place in the sonnets where legal vocabulary and a reference to dream come side by side. That mysteriously lovely one which begins with the prophetic soul of the wide world dreaming on things to come continues immediately with mention of a lease that is supposed forfeit. And when Shakespeare sets about day-dreaming, he 'summons' his memories to his own private court of 'sessions' in the mind.

Alongside this law-court of dream I want now to set another: that which is described in the Barrister's Dream, Fit the Sixth of Lewis Carroll's *The Hunting of the Snark*. Nonsense literature is, I believe, as valid and as closely knit with our ways of thought as any literary genre we have, so this juxtaposition need not, I hope, seem shocking. Its purpose is not to jolt but to help in this investigation, for which these verses provide interesting evidence. Actually, as I have suggested elsewhere, this particular narrative is not pure Nonsense. It admits too much of the real world, which is why it is less successful as Nonsense and highly relevant here and now. Fit the Sixth, like the rest, wavers between flashes of poetry—as in 'There was silence like night' which has a touch of Milton or of Mallarmé, '*Et l'avare Silence et la massive Nuit*'—and an occasional hint of authentic nightmare, 'And the Judge kept explaining the state of the law, in a soft undercurrent of sound.' These are details, however. It is with the Barrister's Dream as a whole that we are concerned, for he dreams, professionally enough, a trial 'in a shadowy Court' where the Snark who is, you may remember, an extremely shadowy entity itself, is Counsel for the Defence on behalf of a pig accused of deserting its sty. The case as it proceeds becomes more and more vague, muddled and self-contradictory. What is interesting is the way in which the various functionaries of the Court abdicate one by one from their functions; the Judge declines to sum up, the Jury refuse to reach a verdict, the Judge cannot pronounce sentence, and little by little the Snark takes on one function after another, returns a verdict of 'Guilty' (although acting supposedly for the defence), and pronounces sentence, 'Transportation for life . . . And *then* to be fined forty pound.' Only at the last is it discovered that the pig had in fact been dead for some years before the case began.

Law-Courts and Dreams

This is not the only law-court in Carroll's dream-writings. *Alice's Adventures in Wonderland*, which is a dream from start to finish, contains two trials, each resembling the more developed law-court in *The Hunting of the Snark* in a number of ways. The trial of the Knave of Hearts at the end of Alice's story proceeds in a no less vague, muddled, topsy-turvy fashion. The jury, lizards, mice and birds as they are, are luckless and incompetent. Witnesses are threatened. The King, sitting as Judge, has no idea of procedure, and due process is subverted—'Sentence first, verdict afterwards.' The second trial in this book occurs earlier and has the look of gratuitous interpolation. This is the Mouse's Long Tale, which runs typographically tail-wise down the page in ever-diminishing print, ending in a whisper as it were: 'I'll be judge, I'll be jury,' said cunning old Fury: 'I'll try the whole cause and condemn you to death.' So ends the trial proposed as a pastime by Fury, the dog, to his mouse-victim. What is interesting is that here, too, Fury, who begins as prosecuting counsel and apparently in fun, absorbs as did the Snark the other functions in the Court, and the trial ends lamentably for the accused.

This gathering up of various judicial roles into one person who then 'embodies the Law' is significant. Surely, the memory says, there is another instance of this in Victorian literature; and then *The Mikado* comes to mind, and Pooh-Bah who was everything from Attorney-General to First Commissioner of Police, taking in Solicitor, Registrar, and Lord Chief Justice on the way, to mention only his legal offices. Here is a pluralist *par excellence*, and the opera is haunted by the Lord High Executioner and moves towards a tribunal and justice and mercy of a sort. Gilbert-and-Sullivan is evidence in its own right about law-courts and dreams, intimately connected with the fabric of the State as these remarkable operas are and as Gilbert Murray attested when he called Gilbert the English Aristophanes. We shall return to Gilbert later, but meantime this dream fusion of legal personnel in Carroll's law-courts is interesting in two other ways. It is, first, an example of those tendencies towards synthesis in dream which Freud wrote of in his great work *The Interpretation of Dreams*, 1900, a process which he called the 'work of condensation'. Second, such a synthesis is no longer a mere matter of the imagination, but has become, within living memory, cold and recurring fact. What were noted or invented as dream phenomena have become actual practice in certain Courts. The metaphor has come true.

Accompanying this metamorphosis of dream into reality, the metaphor or simile or simple conjoining of dream and law-court has continued to appear fairly frequently throughout the nineteenth and twentieth centuries. It took varied forms. There are the two widely different dreams of celestial judgement in Byron and Newman; the extraordinary interplay of dream and law-court in *The Brothers Karamazov*; Stephen Spender's play *Trial of a Judge* is in part a dream-play; one might include *Darkness at Noon*; and I have not forgotten Kafka.[1] For the present I want to stay with the English, however, since they have given us our starting-point, and to consider English justice from the dream point of view.

The odds are that the first any English child knows of the judicial system will be through *Alice in Wonderland* and *Trial by Jury*. My own experience, probably reasonably typical, was that I was well acquainted with Alice by the time I was seven, and with Gilbert and Sullivan since babyhood although with a sense of departed glories since my generation could never hope to compete with the productions achieved by my mother's family, who had been eight in number, four boys, four girls, who could muster among themselves two excellent pianists, two violins, a 'cello, and four professional choristers, not to omit my eldest aunt who had a presence, a contralto voice, and a superbly dramatic blue dressing-gown. As a nine-year-old I was taken, as part of my education, to see the Judge at the Assizes walk in procession to his Court. This one remembers, one's first glimpse of the Law (we were not at all a litigious family): an old man in scarlet, a face like a vulture in a Tenniel illustration, and carrying that prescribed nosegay of flowers which need no longer keep off the smells of the populace but which certainly made a child catch its breath at that visible image of natural innocence in the gripe of some inexorable power. One guessed then, and I believe correctly, that whatever English justice may be, it is not solely rational.

When the layman thinks of the Law and its workings, he tends to envisage it as logic pre-eminent. Dazzled by the law-court's supreme spectacle of dialectic between selected and trained intellects, aware of the absolute precision required, the sharp rules, the sifting of facts and the weighing of evidence, the layman may feel that the practice of the Law is logic through and through, *expertise* of dialectic and deduction made manifest. It is partly this, of course; but 'the fallacy to which I refer is the notion that the only force at work in the development of the law is logic'. So said Judge Oliver Wendell Holmes, in 1897. Does justice then, or at least Anglo-Saxon justice, stretch out a hand to some counter-principle to logic? I have suggested already that it may do so simply by its carefully preserved accretion of ritual, which it shares with English Parliamentary procedure. Ritual is not logic, nor is it a language directed primarily to the reasoning faculty. It is a dream language, and it is to our good dreaming faculty that it speaks wherever it occurs. This is why attacks upon ritual as irrational fly wide of the mark. The very purpose of ritual is to supply the dream content to the mind, to make up the deficient speech of rationalism, however perfect; to raise the *Code Civil*, let us say, to a poetry better befitting the human complexity and fullness of the Body Politic.

It is not merely in its ritual, however, that English justice admits what we may call a dream element into the law-court. It seems to me that it is also admitted, and fundamentally, in that most English institution, the jury.

It is not within my competence to give an outline of the development of

Trial by Jury in this country. For this we may refer to authorities such as Forsyth, from whom we shall discover how controversial such a history is. The institution is claimed, however, as a purely indigenous growth in England, making its way down, from very shadowy beginnings, through the Constitutions of Clarendon, Magna Carta, a very clear statement by a Chancellor Henry VI, one Fortescue, past the turmoil of the Civil War into modern times. It was not, we are told, introduced into France until 1789, into Germany until the mid nineteenth century. It was built into the American Constitution from the beginning.

What comes with the institution of the jury as part of the law-court, inserted into that cross-play of dialectic and that weight of precedent? First, plurality as against the simplicity of a single expert sitting in judgement; though unanimity is required for the jury's verdict, it is unanimity out of difference and not a one-man *pronunciamento*. Second, lack of knowledge of the law; this sounds curiously negative, but the jury's province is fact via evidence, not law, and a jury of lawyers might well be a horror and would certainly be an irrelevance *qua* lawyers. Third, ordinariness; those twelve are to be ordinary fallible people, not appointed by authority nor specially selected nor trained. The jury then introduces into the ordered logical structure of the law-court a counterbalance of plurality, disorder, unreason. It is a remarkable notion. A balance of reason and unreason, of logic and dream, is the very essence of a balanced mind, that is, of sanity. It is also, fundamentally, the operative principle of poetry. There is more than a hint here that justice may also need that same poetic balance. The phrase 'poetic justice' may be something rather more than just a cliché.

If it is conceivable that an element of imagination, of dream whose dynamics, whatever they are, are not those of logic, enters into the law, the converse may also be true. If law-courts are partly dream, then our dreams may also be partly law-courts where we pass judgements on ourselves.

It is a question which seems to have interested the poets in particular. It is always necessary to remind ourselves nowadays, when dreams are under discussion, that neither Freud nor any psychologist since him holds any primacy or monopoly in the study of dreams as valid and meaningful psychic phenomena. The poets have been at this a long time, and it is they who seem to maintain that there is a definite relation between man's dreaming imagination and his conscience, taking a view which is either more strict or more all-inclusive of the human personality than that taken by theology or psychology, both of which absolve us from morals in dreams. Not so the poet. 'In dreams begin responsibilities,' says Yeats, quoting an 'Old Play', and Calderon stands right behind him. The theme of moral judgement passed upon the self by the self in a dream comes out with great clarity and beauty in Grillparzer's *Der Traum ein Leben* where in his dream the hero sees himself, in his passionate ambition, drawn steadily onwards with the remorseless logic of nightmare from a small wrong to a great one and then a host of wrongs, till eventually he metes out to himself the retribution he has merited and wakes in terror. It was only a dream, a hypothetical structure of judgement erected in sleep between nightfall and morning. Yet when the day comes, the dream judgement is upheld as valid and the sleeper remakes his waking life accordingly. In a more modern version of dream-judgement, the soldiers in Christopher Fry's wonderful play *A Sleep of Prisoners* re-enact in communal dream one Biblical scene of judgement after another until in the last episode, still dreaming, they find God's release from the appalling dream-cycle of sin and retribution.

And when seen as a fellow-myth alongside these other speculative constructions by which dreams may be explored, how helpful Freud's Dream Censor is! Freud postulates such a one in Chapter IV of *The Interpretation of Dreams*, 'We should then assume that in every human being there exist, as the primary causes of dream-formation, two psychic forces (tendencies or systems), one of which forms the wish repressed by the dream, while the other exercises a censorship over this dream-wish, thereby enforcing on it a distortion. The question is, what is the nature of the authority of this second agency, by virtue of which it is able to exercise its censorship?' Here the image of a tribunal is built into the formative processes of dream. Freud considers this tribunal generally beneficent, but the image itself is interesting, the more so because Freud when he describes his censor does so in political rather than moral terms. This censor is not the guardian of public morals but of the governmental *status quo*. He holds an office more political than juridical.

The two are, of course, closely connected. Law-courts themselves are political if they are seen in their relation to the legislative functions of the State. Certainly such rights as trial by jury have always been held to have a close connection with democracy and man's fundamental rights thereunder. One of the first things a corrupt government will do is to undermine the inviolability of its courts of law. Examples of this are manifold and contemporary. Where this happens, we may have a commentary on that proper balance of logic and dream which must enter into the making and preserving of public institutions. Where such institutions are perverted and come to grief it may be that the dream functions of the public in question had gone wrong first, starved of their due nourishment and so capable of being supercharged into nightmare. Dame Rebecca West has a remarkable passage on this subject, in connection with Nazi Germany, at the beginning of her book, *A Train of Powder*. Before we come to real life, however, we need to draw what more we can from our literary sources. For this we will go back for a while to Carroll and Gilbert. I said that English children may well get their first notions of the court of law from *Trial by Jury* and *Alice in Wonderland*. It is certainly arguable that, if so, they get rather a queer view of it.

Law-Courts and Dreams

Take Gilbert first, and *Trial by Jury*, of 1875. The scene is laid in Court; the case is one of Breach of Promise; the plaintiff is very pretty. There are songs for each of the participants in the trial, usher, jury, defendant, plaintiff, culminating in the Judge's song, 'When I, my friends, was called to the Bar,' which is Gilbert-and-Sullivan at the top of their form. Meantime instructions are given to the jury in respect of the plaintiff, 'Condole with her distress of mind,' and in respect of the defendant, 'What *he* may say you needn't mind,' with the chorus, 'From bias free of every kind/This trial must be tried.' The jury menace the defendant who expostulates, 'These are very strange proceedings,/For permit me to remark,/On the merits of my pleadings/You're entirely in the dark.' He tells his story, to which the jury reply that they too were fickle in their younger days but 'I'm now a respectable chap,' they declare, 'And shine with a virtue transcendent,/And therefore I haven't a scrap/Of sympathy with the defendant.' The Judge, no less candidly, relates his escapade of ensnaring and then jilting the Old Attorney's Elderly Ugly (Wealthy) Daughter—'And now, if you please, I'm ready to try/This case of Promise of Marriage.' In the end justice is done by the Judge deciding to marry Angelina himself.

This tiny opera is very simple, and sweet, fun. It exhibits the all-too-human weaknesses of any judicial system which admits the human element in place of a mechanical excellence, but is essentially a light-hearted commentary on a precept which could undo all human law-proceedings from their inception: 'Let him that is without sin among you cast the first stone at her.' It is absurd, the playlet says, for sinners to sit in judgment on their fellow-sinners, and the outcome is inevitably irrelevant, but in this case cheerful.

This is not the only Savoy Opera to deal with the Law. *Iolanthe*, of 1882, also does so, and here a dream-like element is included from the beginning, in the relationship between the House of Peers, who are the supreme Court of Appeal in English Law and so uniquely combine the functions of the legislative and the judiciary, and Fairyland; the Lord Chancellor who embodies the Law and is of so susceptible a heart and the formidable Queen of the Fairies. It is the Chancellor who sings the famous Nightmare patter song, and also the Chancellor who at the end of the play proposes a solution to the problem created by a fairy law decreeing that every fairy who marries a mortal shall die. 'Allow me,' he says, 'as an old equity draughtsman, to make a suggestion. The subtleties of the legal mind are equal to the occasion. The thing is really quite simple—the insertion of a single word will do it.' That word is, of course, the word 'not'.

This is curiously, and perilously, like an incident in the trial at the end of *Alice in Wonderland*. Alice has been called to take the stand as a witness, and has averred that she knows nothing whatever about the case. The king, who is presiding Justice, then says, 'That's very important.' He is prompted to amend this to 'Unimportant', after which he goes on murmuring, 'Important—unimportant—unimportant—important,' as if, the narrative says, to see which sounded better. Where the other two trials of Carroll's emphasize the running together of roles and the condemnation of the prisoner, the main note of this one is a crescendo of hypnotizing incoherence, of which this equivalence of importance and unimportance is part, approaching closer and closer to nightmare until Alice breaks out of the dream before the trial is finished. There is a general foreshortening of procedure: the jury are to consider their verdict immediately after the accusation has been read; and there are touches of nightmare, for instance the casualness with which the Queen of Hearts, seeing a witness leaving the box, says, 'And just take off his head outside.'

This dream trial was composed in 1865. In 1938 Stephen Spender published his *Trial of a Judge*. Two acts of this play are dreams, the first in the mind of a judge who has just condemned to death, for a political murder, three Fascists, in a country on the brink of a Fascist revolution, the second in the mind of a Home Secretary who has to see that sentence rescinded and three Communists substituted for the three Fascists.

Between Alice and Iolanthe on the one hand and Spender's Judge on the other, there lies a series of remarkable trials. Dreyfus in 1894; the Reichstag Fire in 1933; the Purge Trials in Moscow in 1937 and 1938. The series did not end there, for the McCarthy investigations in the United States after the last war were of the same order. These were all trials where politics of a highly specialized kind obtruded into the law-courts, and strange things happened. Those involved in these trials, as victims or observers, recognized the precedents[2] and repeatedly characterize the proceedings as dream, nightmare, fairy-tale, or *Alice in Wonderland*.

A number of examples follow, chosen not for political or literary merit but simply to show the frequency of this insight. Frederick Birchall in *The Storm Breaks* says of the period between the Reichstag fire and the trial, 'the utterly fantastic was attested by showers of affidavits, and by contrast with some of the official communiqués Grimm's *Fairy Tales* became narratives of sober fact.'[3] Victor Kravchenko in *I Chose Freedom*, speaking of the Russian trials, says, 'For seven days entire pages in the press detailed the nightmarish "confessions" of these men.'[4] Later he speaks of 'the N.K.V.D. fairy-tales',[5] both phrases recurring elsewhere in the book. George Marion, in *The Communist Trial: An American Crossroads*, applies to Judge Harold Medina's handling of the 1948-9 trials of American Communists a variant of the Wonderland phrase: 'Sentence first, trial afterward',[6] uses *Harold in Wonderland* and *The Law in Wonderland* as chapter titles, and quotes, 'I'll be judge, I'll be jury',[7] besides comparing proceedings to the Reichstag fire trial. (H. R. Trevor-Roper, incidentally, in *The Last Days of Hitler* compare the meeting of the Nazi leaders after the bomb plot on Hitler's life to the Mad Hatter's Tea-Party.)[8] Richard B. Morris in *Fair Trial*, dealing with the Hiss case, relates that Hiss's counsel said to the accused at the close of his final speech, 'Alger Hiss, this long nightmare is drawing to a close';[9] Morris later compares the Hiss Trial to that of Dreyfus, if it be supposed that Hiss was innocent.[10] Finally, James A. Wechsler, in *The Age of Suspicion*, says of his own investigation by McCarthy, 'At least this story may underline the nightmare quality of some of the inquiries to which this country is now being subjected,'[11] later repeating the nightmare image[12] says, 'There were moments during the interrogation when I thought of the Moscow trials, and what it must have been like to be a defendant,'[13] asks, 'How could one break through the ring of fantasy that McCarthy was constructing?'[14] and describes the concluding moments of this tribunal by saying of McCarthy himself, 'He was very much the judge now, handing down the decision in favour of the prosecutor (who happened to be himself).'[15] Here are Snark and Old Fury in grim earnest. The last logical step is taken by Kafka's character, Joseph K., in *The Trial*: 'K. now perceived clearly that he was supposed to seize the knife himself as it travelled from hand to hand above him, and plunge it into his own breast. But he did not do so ... He could not completely rise to the occasion, he could not relieve the officials of all their tasks.' He is executed none the less. And is this literature, nonsense, nightmare, or sober and well-attested fact?

The actual cases I have cited occurred in France, Germany, Russia, and the United States, but it is strange that a blueprint for what happened should be found in English nonsense literature many years earlier, and it should at least make us realize our involvement in what has gone amiss elsewhere in the balance between reason and dream, precarious enough, on which all good government, of the public weal or the private mind, depends.

On the whole it is the poets, particularly the dramatists, who have worked at this point of interchange between private and public forms, more particularly the dream forms which in their turn must build up the body of the State. How deeply they have worked at this one aspect of their many-sided task, dreaming of law-courts, can be glimpsed in the great vision of a remaking of celestial justice, aided by a human court of law, in *The Eumenides*, or in that hallucinatory tribunal set up by the mad King Lear, his justicers the disguised Kent, the feigned madman Edgar, and the Fool, in a scene which is at one and the same time profound human reality, dream, and apocalypse. Here great minds in the law work also, one of whom Judge Cardozo, said of that Law, 'The process in its highest reaches is not discovery but creation.'[16] Spender in his play echoes something akin to this when he says,

> Yet this Judge in the last analysis believed
> That an argument would govern the State which drew its form
> From the same sources as the symmetry of music
> Or the most sensitive arrangement of poetic words
> Or the ultimate purification of a Day of Judgement.

With that in mind I want to end with three suggestions. First, the forms of the body politic, be they legislative, judicial or administrative, are an image also of the forms of the mind, individual minds making up that corporate mind to which we give rather general names such as public opinion or national consciousness or group psychology, of whose nature and possible distempers we know little as yet. Second, the poet is, by virtue of his art which is private and public, actively at work here, to preserve the balance of reason and unreason and to keep us from those sicknesses of both which have deformed our governments and our minds this last century. Third, if dream and law-court commingle as it seems they may, this is only part of a larger process by which the power of dream helps to shape our political institutions and hence our future; and that leaves much room for exploration.

NOTES

1. The similarities between *The Trial* and the two Alice books are discussed by A. E. Dyson in *The Twentieth Century*, Vol. CLX, July 1956, pp. 49–64, in an article entitled 'Trial by Enigma.' I am indebted to Professor Polanyi for calling this to my attention.
2. Judge Learned Hand, speaking in 1952 of precedents for the American phenomena, suggests the England of Titus Oates and the French Revolution, and America after the Civil War and the First War, 'except that in our case,' he adds, 'we have outdone our precedents.' *The Spirit of Liberty: Papers and Addresses*, New York, Knopf, 1952, p. 256.
3. F. BIRCHALL, *The Storm Breaks*, New York, Viking Press, 1940, p. 124.
4. V. KRAVCHENKO, *I Chose Freedom*, New York, Scribner, 1946, p. 234.
5. ibid., p. 235.
6. G. MARION, *The Communist Trial: An American Crossroads*, New York, Fairplay Publishers, 1950, Chapter 6.
7. ibid., p. 68.
8. H. R. TREVOR-ROPER, *The Last Days of Hitler*, New York, Macmillan, 1947, p. 33.
9. R. B. MORRIS, *Fair Trial*, New York, Knopf, 1952, p. 463.
10. ibid., p. 478.
11. J. A. WECHSLER, *The Age of Suspicion*, New York, Random House, 1953, p. 9.
12. ibid., p. 311.
13. ibid., p. 280.
14. loc. cit.
15. ibid., p. 287.
16. G. S. HELLMAN, *Benjamin N. Cardozo: American Judge*, New York, Whittlesey House, McGraw-Hill, 1940, p. 95.[19]

PART FOUR

The Knowledge of Living Things

Thus, at the confluence of biology and philosophical self-accrediting, man stands rooted in his calling under a firmament of truth and greatness. Its teachings are the idiom of his thought: the voice by which he commands himself to satisfy his intellectual standards. Its commands harness his powers to the exercise of his responsibilities. It binds him to abiding purposes, and grants him power and freedom to defend them. *P.K.*, p. 380

17

The Logic of Biology

MARJORIE GRENE

1

'The ontology of commitment . . . can be expanded by acknowledging the achievements of other living beings. This is biology. It is a participation of the biologist in various levels of commitment of other organisms, usually lower than himself. At these levels he acknowledges trueness to type, equipotentiality, operational principles, drives, perception, and animal intelligence, according to standards accepted by him for the organisms in question. . . . These achievements are personal facts which are dissolved by any attempt to specify them in impersonal (or not sufficiently personal) terms. The unspecifiability of such achievements can now be seen to represent a generalization of the theorem that the elements of a commitment cannot be defined in non-committal terms. The paradox of self-set standards and the solution of this paradox are thus generalized to include the standards which we set ourselves in appraising other organisms and attribute to them as proper to them. We may say that this generalization of the universal pole of commitment acknowledges the whole range of being which we attribute to organisms at ascending levels.'[1]

This essay may be taken as a commentary on the above text, or, more generally, on the argument of Part Four of *Personal Knowledge*. I am trying to elucidate Polanyi's conception of biology by stating, very simply, some aspects of it and interpreting some assertions of biologists in the light of this statement.

Epistemologists sometimes distinguish between knowledge as recognition (as when I know a friend when I see him) and knowledge that a proposition is true (as when I know that mammals have four-chambered hearts). The theme of *Personal Knowledge* may be said to contradict this distinction, in so far as it stresses the element of recognition which is essential to all acts of knowing. I want to consider some of the implications of this assertion—that recognition is an essential ingredient of all knowledge—for our understanding of biological knowledge. This may also suggest some of the more general philosophical problems on which the argument of *Personal Knowledge* sheds new light.

Let me take as my starting-point a paper by Professor C. F. A. Pantin on 'The Recognition of Species', in which he contrasts the yes-no matching of specimens against characters by museum taxonomists with the informal 'aesthetic' recognition of species in the field.[2] Professor Pantin is chiefly comparing a deductive process: All such-and-suches and only such-and-suches have characters 1, 2, 3, 4; specimen n has characters 1, 2, 3, 4, therefore specimen n is a such and such, with the intuitive recognition of an individual as belonging to a certain kind: as when he tells his students to bring in 'all the worms that sneer at you', or when he finds a specimen of a new species and exclaims, 'Why, it's a rhynchodemus but it's not bilineatus, it's an entirely new species!' But he concludes his essay with the suggestion that inductive as well as deductive inference has traditionally been treated as if it were of the same yes-no character, and he tells us that we ought to consider whether 'aesthetic' elements may also be implicated in the inductive as well as the deductive processes not only of biology, but of science as a whole.

Now Professor Pantin's 'aesthetic' elements in science seem to me to be identical with what Polanyi calls its *unspecifiable* components.[3] What I want to do here is to look at these aesthetic or unspecifiable constituents both outside and inside biology and see how biology both differs from and resembles the exact sciences in its reliance on such constituents. By unspecifiable constituents, it should be noticed, Polanyi does not mean propositions that are held to be probable rather than true or false, nor arguments that yield probability rather than truth or falsity. The probabilities which induction is supposed to yield, as Professor Pantin also points out, are intended to be just as yes-no as are the certainties of mathematics. I shall return to this problem briefly later on. The point here is threefold: (i) There are constituents of knowledge which, though not only psychologically but epistemologically indispensable to it, are not stateable in the form of propositions or arguments. That is what is meant by calling them *unspecifiable*. (ii) Such constituents of knowledge are unspecifiable because they are *personal*. They exist because knowing always expresses a personal commitment, and a commitment can never be wholly reduced to, or exhaustively stated in, non-committal form. (iii) Knowing always expresses a personal commitment, because it entails the apprehension of a whole in terms of its parts, or of an aim in terms of the means to it. It entails, in Polanyi's language, both *focal* and *subsidiary* awareness.[4] Although I may, through analysis, bring into focus aspects of my knowledge which were subsidiary, and allow to lapse into subsidiary knowledge what I used to focus on, wholly subsidiary knowledge would be simply forgotten knowledge and wholly focal knowledge would be trivial. It is the fusion of the two into one which constitutes, so to speak, the flesh and blood of knowledge, which definitively shapes my act of knowing what I am knowing whenever I am knowing it. Such fusion is essentially, necessarily, logically a personal act, and no formalism, however ingenious, can make it otherwise.

2

Accepting this general conception of unspecifiable constituents in scientific knowledge, I shall try to distinguish four types—or better, four levels, at which they occur, for they seem to form a hierarchy, in so far as each entails all the previous members of the quartet. Adopting my nomenclature from Professor Pantin's suggestion about 'aesthetic recognition', I shall call the first type the recognition of pattern, the second the recognition of individuals, third, recognition of persons, and fourth, recognition of responsible persons.

The first is common to the biological and the physical sciences—or for that matter to any knowledge at all. What I am referring to here is often called an awareness of Gestalt—and a familiar example in the history of science is Kekulé's reported day-dream of snakes chasing their tails which is supposed to have led him to the discovery of the benzine ring.[5] All discovery relies in the last analysis on such intuitive perceptions of form. The achievement of the scientist in successful pursuit of a new theory is to glimpse a Gestalt as yet unseen by his predecessors or contemporaries. Maxwell's equations are a paradigmatic example of this. As Max Born says in his *Theory and Experiment in Physics*, 'Maxwell's addition of the missing term is just such a smoothing out of a roughness of a shape.' And of such processes in general he says: 'A synthetic prediction is based on the hypothetical statement that the real shape of a partly known phenomenon differs from what it appears to be.'[6] And the testing of a theory, then, is the endeavour to find out by accepted procedures, mathematical or experimental, whether the Gestalt thus envisaged is 'really' there. Nor are these testing procedures necessarily yes-no; they may be themselves unspecifiable or aesthetic. The theory of relativity, Professor Dirac has said, was accepted for two reasons, its agreement with experiment and the fact that 'there is a beautiful mathematical theory underlying it, which gives it a strong emotional appeal', and of these the latter reason, in his opinion, was the more important. 'With all the violent changes to which physical theory is subjected in modern times,' Dirac writes, 'there is just one rock which weathers every storm, to which one can always hold fast—the assumption that the fundamental laws of nature correspond to a beautiful mathematical theory. This means a theory based on simple mathematical concepts that fit together in an elegant way, so that one has pleasure in working with it.'[7] What is often misleadingly referred to as the 'simplicity' of theories is an aspect of this aesthetic component: a theory *feels* simple when the mind rests happily in the pattern it offers.[8] This is not of course by any means always a pattern in the visual sense of Kekulé's snakes; it may be an intellectual pattern—a mathematically elegant formulation, a formulation which as Dirac says mathematical physicists like working with, or, in biology, statistical generalizations like those of population genetics which geneticists like to work with. Or it may be a model from a familiar aspect of our experience applied analogically to the unfamiliar, as in Darwin's application of the experience of stockbreeders and pigeon fanciers to the origin of species. (There is, of course, much more to Darwin's argument than that—that's not my subject here; but Darwin's theory does have an astonishing way of bringing the remote past into the area of the familiar and making us feel that we know how it was because it *was* the way it still *is*.)

I have suggested that we call this first kind of unspecifiable factor in knowledge the recognition of pattern; perhaps we might describe it even more weakly and inclusively as a sense of relevance; and its necessary presence in this minimal form becomes evident if we look, as we may do here briefly, at one of the many recent attempts to analyse the structure of scientific knowledge without admitting such a factor. Consider for example Sir Harold Jeffreys's *Scientific Inference*.[9] Jeffreys is, in effect, describing scientific inference as the reiterated application of Bayes's theorem of inverse probability. But Bayes's theorem starts from an initial probability. How do we obtain this? What his theory amounts to, Jeffreys says, is that he has turned the traditional principle of causality upside down: we start, he says, with *random* correlations and work in the direction of necessary connections as our objective.[10] Yet, as Bayes's theorem indicates, random events are precisely what we do *not* start with, for they would be random only if they did not entail any prior probability, or conversely, if they were random we could not establish any initial probability and so could never begin. Only the acceptance of a rational context, the choice of one set of data as more relevant than another, will get us started at all. Bayes's theorem formalizes a procedure applicable within an accepted context, but not the discovery of context, the recognition (if we may platonically call it so) of novel pattern, and with it scientific advance, inference that is heuristic rather than routine. In other words, it is the recognition of pattern that supports the logical gap between evidence and theory.

Incidentally, I might bring similar objections against Professor Braithwaite's rules of rejection in his *Scientific Explanation*: for one must choose such a rule, and one's reason for choosing depends upon a sense of relevance, or of irrelevance.[11]

3

The recognition of pattern, then, is essential to all scientific discovery, and at one remove therefore to the mastery of any scientific discipline by the student—which is a process of discovery for him though not for humanity as a whole. This much is true of biology and of the exact sciences equally. But in addition to this common situation there is also a difference. Once recognized, an explicative context in the exact sciences can be left far behind in its routine application. In biological practice, on the other hand, in the field recognition of which Professor Pantin is speaking, or equally, for example, in medical diagnosis, the awareness of pattern continues all along to play a prominent part. Nor is this because biology is a 'younger' science or a less 'developed' science, but simply because it is *biological* science.

This is a tautology, but it is a heretical one, for it is an article of faith with many, if not most, biologists, that their science is really not biological at all but is only physics and chemistry writ large. And when they get the writing small and precise enough, they say, it *will* be physics and chemistry. So, to take one example of many, say Fraenkel and Gunn in their classic work on *The Orientation of Animals*.[12] And yet the kind of skill described by Professor Pantin as aesthetic recognition persists in biology. Is this then a remnant to be superseded when mathematical biophysics takes over? I think not; for the practice of biology entails a recognition of pattern in a more pervasive sense than do physics and chemistry. Over and above the recognition of abstract patterns characteristic of the sciences of inanimate matter, the practice of biology demands the recognition of individual living things, and analysis in biology is always analysis *within* the context set by the existence of such individual living things. Thus a second kind of aesthetic recognition, the recognition of individuals, adds to the subject matter of biology a logical level missing in the exact sciences, and at the same time limits the range of analysis to the bounds set by the acknowledgment that individual living things exist. I do *not* mean that at some mysterious point analysis will have to stop, but that an analysis of an organism which analysed the organism *away* would contradict itself by destroying its own subject matter. Nor do I mean that when we recognize an individual we are adding some mysterious vital something that comes from I know not where, but that we *are* affirming the existence of something which is more than a brute fact, in the sense that we acknowledge it as an achievement: as an entity that succeeds or fails relatively to standards which we set for it. It is a good or a bad specimen of *Cepaea nemoralis* or *Spiraea vanhoutiens*. We recognize it as an individual in respect to its trueness to type;[13] and no matter how far analysis may proceed, this recognition will always be essential. Otherwise we should not know what we were analysing.

Again, I have to lay this down here as a flat pronouncement against the authority of the biological profession itself, or a large and authoritative part of it; but let me try to support my assertion by reference to one of the branches of biology which considers itself well on the way to the ideal of withering away into biochemistry. As everyone knows, genetics since the revival of Mendel's work at the beginning of this century, and since the work of Johannsen, Morgan, and many others, has been founded on atomistic principles. Its guiding maxim seems to be: if we could specify all the genes we could specify the organism. And the brilliant work of recent years on the chemistry of the cell nucleus and on reproduction in viruses and bacteria phage has given this maxim new and apparently overwhelming support. Thus a leading geneticist, summarizing this work at a recent conference for teachers of biology, spoke of the D.N.A., R.N.A. and protein composing the nucleus and set as the theme of the experiments he was reporting the question: 'Which of these is the genetic material which chiefly causes the effects that we see?'[14] This question expresses for biology the same kind of hope that Henry Oldenburg, the first secretary of the Royal Society (as he was to be) expressed for natural philosophy as a whole when he wrote to Spinoza: 'In our Philosophical Society we indulge, as far as our powers allow, in diligently making experiments and observations, and we spend much time in preparing a History of the Mechanical Arts, feeling certain that the forms and qualities of things can best be explained by the principles of Mechanics, and that all the effects of Nature are produced by motion, figure, texture, and the varying combinations of these.'[15] Divide and conquer! Specify the parts and you have the whole. The parts of an organism are chemical molecules; specify these and you need worry about 'life' no longer.

But parts by definition are *of* a whole; and as genetical research proceeds, along with specification, the nature of the whole, too, makes itself felt. The parts are the *conditions* for the whole, which certainly could not exist suspended in some heaven of essences without them; but it is the whole that *explains* the parts, not the parts the whole. The whole is the system (the organism) that makes the parts the parts that they are, even though the parts are the conditions (in traditional language, the material causes) for the existence of the whole.

So far this is only to say that all explanation is systematic, and this we have admitted already in different words in saying that all discovery entails awareness of form or Gestalt. But biological explanation—and that is my point in reference to genetics—entails the recognition not only of systematic connections—between such genes and such phenotypes—but of individually existent systems: organisms existing as unitary four-dimensional wholes, as individuals with a life history in a particular portion of space-time. This ought to be clear from the breakdown of one-gene-one-character genetics, but the physico-chemical, atomistic habit of thought is so strong among biologists that few of them recognize the re-orientation which this change implies. It is admitted explicitly, however, in a very strong statement by that eminent elder statesman of genetics, R. B. Goldschmidt, in his *Theoretical Genetics* (and Goldschmidt is certainly no 'holist' or 'vitalist' or anything of the sort, but an old-fashioned materialist who has the honesty and courage to admit what confronts him). This is what he says at the beginning of his account of the action of the genetic material:

'Here, at the start of our discussion of genic action, one point should be made clear. It is one of the general tenets of genetics that a mutant locus of the gene, assumed to be the normal allele, does not control a character but is only a differential: the visible character depends upon a large number of genes,

The Logic of Biology

if not on all of them. This idea is frequently illustrated by the fact that many loci are known to influence the same character if mutated. Thus the numerous eye-colour mutants in *Drosophila*, scattered over all chromosomes, would indicate that there are at least that many genes for eye colour. The work on biochemical genetics of eye colours as well as nutritional requirements in *Neurospora* shows that a number of mutant loci individually interfere with different steps of organic synthesis from the lowest raw material: for example, in eye-colour synthesis, from trytophane to kynurenine, then to 3-hydroxykynurenine and further steps not yet well known. A corollary is that for each of these steps a number of mutant loci are known, which interfere with it specifically. If we wish to express this factual situation by saying that a phenotypic trait is the product of action of many or all genes, we must realize that this *façon à parler* is *nothing but a circumscription, in terms of the atomistic theory of the gene, of the unity and integration of the organism.*'[16]

Now notice that the atomism is here supplied by the theory and it is the unity that is the fact: that is the important point. In scientific discovery, as common to biological and physical science, the unity is the new context, the theory; the facts may be all over the place—as are the fossils, the embryological structures, the pigeons, the clover, field mice or old maids relevant to the Darwinian theory of evolution; or as are the scattered populations of Neurospora, Drosophila, castor beans or cattle relevant to the atomistic theory of genetics. But in biology, over and above this unifying relation of all theories to their data, there is the unity of the individual living thing underlying the abstractions of the theory, the unity that *is* the fact on which all the abstractions, atomizing or otherwise, bear.

But surely, it will be objected, theories explain facts. If the theories specify parts, as in genetics (or original conditions as in evolutionary theory), then is it not, after all, the parts that 'explain' the whole? Here the ambiguity of 'explanation' and the prestige of physical theory combine to confuse matters. Ideally, or (what is considered the same thing) in the exact sciences, an explanation provides some sort of formulation, whether a model or a mathematical formulation, from which statements about a certain range of phenomena can be deduced. From theory T we can deduce statements of fact a, b, c, d . . . For even though evidence never entails the theory that explains it, the theory once envisaged entails the evidence. What Goldschmidt's statement tells us, however, is that genetical explanation is never wholly of this sort, since we must supplement our theory by reference to a fact which we can indeed circumscribe in the language of the theory but could never have predicted from it: the fact of the existence of the organism O, not as an aggregate of any number of genes but as an integrated whole. From his breeding experiments the geneticist can indeed predict an impressive range of mutations a, b, c, d . . . n, but no specification of D.N.A's, tryptophanes, or what you will *entails* a, b, c, d . . . n, *except in the context of O*. The recognition of O puts the *facts* of biology in a different and more complex relation to its theories than is the case for the exact sciences. The geneticist's recognition of a fruit fly stands in a different logical or epistemological relation to the theories of genetics from the relation, say, of the reading of the temperature or pressure of a gas to the kinetic theory of gases. For over and above the recognition of pattern implicit in the grasp of data relevant to theories, biology demands the recognition of *individuals*, to which as its *raison d'être* it has continually to return.

I suspect that there is a relation here to the role of causal explanation in the biological as against the exact sciences. Explanation in genetics, as in all the most 'advanced' branches of biology, aims at being 'scientific' by being causal. Yet the exact sciences, we are sometimes told, are not for the most part interested in causes at all.[17] Thus the kinetic theory of gases explains the behaviour of gases under changes of pressure and temperature in the sense that the phenomena follow logically (not temporally) from the theory. But there is no question of before and after and so none of cause and effect. The theory supplies logical reasons, not historical causes. But the sum total of genes times environmental influences is supposed to *cause* the organism. D.N.A. experimenters are seeking for the genetic material which '*causes*' the effects that we see; the story of template reproduction is a *causal* story. What does all this adds up to? To the fact that certain aspects of the organism are controlled by certain conditions, so long as there exists an organism to be affected by such conditions—somewhat as a machine which operates according to a certain principle can be made to work faster or slower, with more or less efficiency, by a number of contributory causes. Like explanations in engineering, explanations in biology depend on the pre-existence and continued existence of a particular whole or a set of particular wholes, machines or organisms respectively, of a particular character. The machine is a whole existing for some end beyond itself, but the organism—though in an evolutionary context existing as a means to the existence of future organisms—is as an individual organism an end in itself: it *is* the system on which causal explanations are based and to which they have to return. In both cases—machines and organisms—the causes specified are relevant to the reasons or principles in virtue of which the wholes in questions exist: in one case the operational principle of the machine, in the other the 'unity and integration of the organism'. And in both cases the causal analysis is possible precisely because the whole—the system being analysed—*is* a historically existent whole, a four-dimensional entity, and not merely an abstraction from which statements about phenomena are deducible *sub specie eternitatis*. Thus the conspicuous use of causal explanation in biology appears to be a sign that biology is 'scientific' not in the same way as, but in a different way from physics and chemistry: that instead of deducing particulars from abstract systems it specifies causal connections bearing on concrete, existent systems, on individuals.[18]

The Logic of Biology

4

To many biologists such statements as I have just been making are either nonsensical or false; they 'smack of teleology' or of 'vitalism'. For if we insist that in explaining organic phenomena scientists are doing something different from what they do in explaining the phenomena of the inorganic world, we are supporting in effect W. D. Elsasser's thesis that there are 'biotonic laws', regularities inexplicable by the laws of physics and chemistry,[19] and so we are re-introducing into the natural order the very discontinuity which Darwin and his heirs had so brilliantly banished. For the fundamental law of biology for a hundred years now has been the principle of *uniformity* which Lyell had applied in 1830 to the rocks alone and which Darwin took to its logical conclusion in its application to the history of life as well. If, as we must surely admit, there was a time when there was no life on this planet, and life has its own laws, then the law of nature are not constant and uniform. And if in understanding and explaining organic phenomena we are proceeding in a fashion different from the way in which we proceed when explaining physical phenomena, this suggests also that *what* we are explaining is in fact existentially and historically different.

Yet to insist on epistemological and even ontological discontinuity is not to deny historical continuity, for conditions which are continuous can give rise to, or trigger, systems which once in existence are self-sustaining and hence not explicable entirely in terms of the conditions which produced them. As it is when we strike a match and produce a flame, so was it when the open systems which are organisms were produced from the conditions prevailing on the earth's surface at the first emergence of life—and I am *not* saying that anyone was there to strike the match, only that the principles sustaining an open system are different from and additional to the conditions necessary to initiate it. This is the point Leibniz was making when he used the analogy with algebra to show how the principle of continuity could apply throughout nature—for if you moved one focus of an ellipse continuously, at infinity you would get a parabola, and so you have a second figure which differs discontinuously from the first yet is produced from it by a continuous process.[20] For the epistemology of biological knowledge, this is an extremely important analogy. The discontinuity of emergence is not a denial of continuity but its product under certain conditions.

5

Once we have admitted the principle just stated, we may grant a unique epistemological status to the recognition of individuals without violating our underlying belief in the continuity of nature. Moreover, I need this principle here also to establish the third kind of aesthetic recognition I want to talk about. Let me return for a moment to Professor Pantin's worms that sneer at you. All recognition of individuals has the same yes-no character whether the individual in question belong to the species *Rhynchodemus bilineatus* or *Ranunculus bulbosus* or *Felis domestica* or *Homo sapiens*. Is there any difference in all these cases? Jennings, the great Protozoologist, insisted that if paramecia were the size of dogs we should know them personally as we do our domestic animals.[21] They are not so and most of us do not know them in this personal way. But still there is a difference between recognizing a buttercup and recognizing an animal that we do know personally. Somewhere along the evolutionary sequence we find animals whom we do recognize as persons rather than simply as individuals of a species. Here we seem to have a second discontinuity in the continuous series of advancing forms of life. Polanyi identifies this division with the distinction between our morphological judgments of trueness to type, as in taxonomy, and our judgments of a more active form of achievement, of success or failure according to rules of rightness, as in the analysis of animal behaviour. He mentions the case of a rat who drinks saccharine solution, 'mistaking' it for 'food'. Such a rat has made what he calls a 'reasonable error'.[22] But buttercups do not seem to make mistakes, nor to succeed.[22a] Not that rats 'think' in these terms—that is not the point; but we do, and we judge the rat's behaviour according to the standards we set for it, according to what seems reasonable for a rat. This is to think of the rat as a centre of appetites and interests which may be satisfied or not, not as an individual simply, but as a sentient individual, that is in some sense as a person. Again, such recognition of incipient personhood establishes a new level over and above the others we have been talking of: to recognize a person means to recognize a coherent form, an individual living thing, *and* a centre of appetites and interests, capable of error and therefore in some degree of rational performance (and if this seems absurd, consider that rats can go mad).

Once more, however, such statements seem to contradict the most cherished maxims of biology. The very existence of the science of ethology, for example, is said to depend on the resolute renunciation of any ascription to animals of 'sentience', feelings, or the like. And indeed the scientific study of animal behaviour could not have got far through accounts like those of Romanes in the eighties. In comparing a bird's flying into a window with insects flying into a flame he says:

> 'Here there can be no question about a possible mistaking of a flame for white flowers, etc., and therefore the habit must be set down to mere curiosity or desire to examine a new object; and that the same explanation may be given in the case of insects seems not improbable, seeing that it must certainly be resorted to in the case of fish, which are likewise attracted by the light of a lantern.'[23]

This does indeed seem a strange misinterpretation of instinctive behaviour, and by contrast one might reasonably suppose that only a strictly 'objective' account of what the animal in fact does and an objective analysis of the factors 'eliciting' its responses could produce a science of ethology. So e.g. argues Dr. Tinbergen in opposition to an animal psychologist like Bierens de Haan.[24] But again it seems to me that (though no one could wish to reinstate Romanes's indiscriminate subjectivism) the magnificent analytical achievements of modern behaviour studies do in fact take place within the context of the recognition of persons. Again, let me take one example to illustrate this: Tinbergen's popular but nevertheless detailed and systematically 'objectivist' *Herring Gull's World*.[25]

That instinctive behaviour lends itself in large measure to objective analysis is plain. Take the case of egg retrieving.[26] The gull doesn't 'think' of putting out its wing and sweeping the egg back in to the nest, but executes what look to the bird-watcher like a succession of clumsy and ineffective movements. Clearly its behaviour is determined by a series of rigid and precise instinctive mechanisms. It would be quite irrelevant to try to imagine what the gull was 'thinking' about in such cases. The whole apparatus of Internal Release Mechanisms, for instance, in the intricate series constituting nesting behaviour,[27] works so much more rigidly than our so-called intelligent appetitive behaviour that analogies with human activities are more often than not misleading; so the maxim of objectivity is methodologically and factually correct up to a point.

At the same time the ideal of objectivity repeatedly breaks down in Tinbergen's account. For one thing, there are his references to the 'meaning' of behaviour; e.g. the upright threat posture is 'full of meaning' for all herring gulls,[28] or, more subjectively, the choking situation is 'really loaded with hostility'.[29] Sometimes 'meaning' clearly refers to functional significance such as could be explained by natural selection—for example when Tinbergen reports, 'The bird makes some curious motions with the bill without making any sound. . . . I do not know what these motions mean.'[30] Here to find the meaning might be to find the function in terms of the evolutionary conception of the bird as a machine for the survival of its descendants. But the references to threat posture or choking are not of this type.[31] In general, moreover, such analyses suggest at the least that behaviour patterns are not being simply described, nor yet reduced to physico-chemical terms, but referred to in terms of an understanding of the whole life of the organism or the community of organisms. For instance, Tinbergen describes as follows a piece of behaviour difficult to interpret:

'One bird walks round the other, uttering a peculiar call not unlike the begging call of a half-grown Herring Gull chick. It tosses its head up and down and even touches the other bird's bill. The latter seems to try to get away, but the first one keeps bothering it. Finally, the harassed bird stands still, twists and turns its neck, a huge swelling appears in its neck, moves upward, and suddenly the bird bends its head down, opens the bill widely, and regurgitates an enormous fish. The begging bird begins to eat gluttonously before the food even reaches the ground. The other joins it, and together they finish the meal. It is easy to understand that this behaviour was feeding. *But who was feeding whom, and why?*'[32]

This is far from the description in terms of muscle contractions which Tinbergen had earlier set up as the ethologist's ideal.

In this case, moreover, we have a clear instance not only of the degree of personhood implicit in appetitive behaviour in the individual, but of interpersonal relations also. And elsewhere Tinbergen speaks more generally of 'genuine personal ties',[33] and says that 'personal likings and dislikings play a prominent part' in gull life.[34] Elsewhere also he points out that gulls know the difference between the alarm calls of 'nervous, panicky' gulls as against more placid individuals, since they fail to react to the cries of the former.[35]

Further, not only the gulls, but Tinbergen himself came to know the gulls apart: 'It is quite a thrill', he writes,

'to discover that the birds you are studying are not simply specimens of the species *Larus argentatus* but that they are personal acquaintances. . . . Somehow, you feel, you are at home, you are taking part in their lives, and their adventures become part of your life. It is difficult to explain this more fully, but I think everybody who has studied animal communities will understand how we felt.'[36]

This is surely a striking admission by a leading programmatic objectivist of the recognition of persons—and *a fortiori* of an unspecifiable constituent of his science, 'which is difficult to explain but which everyone who has studied animal communities will understand.'

And finally, although the interpretation of gull behaviour in terms of human subjectivity is rejected, the interpretation of *human* behaviour in terms of *gull* behaviour is not. 'Much of what little understanding I have of human nature,' Tinbergen confesses,

'has been derived not only from man-watching, but from bird-watching and fish-watching as well. It is as if the animals are continuously holding a mirror in front of the observer, and it must be said that the reflection, if properly understood, is often rather embarrassing.'[37]

In short, Tinbergen, for all his objectivistic faith, gives overwhelmingly the impression of a man who knows not so much the physics of muscle contractions or the chemistry of nuclear proteins, as *sea-gulls*, and that in a personal way that is different from knowing physical or chemical phenomena or even from knowing buttercups and worms.

6

What, in conclusion, about my last type, or level, of aesthetic recognition? This takes us beyond biology, but in a direction suggested by the road we have been travelling so far. In the last passage I have quoted from Tinbergen, the ethologist puts himself in a sense on the same level, or even on an inferior plane, to the herring gulls; but in general his comparisons, for instance between gulls and crows or between gulls and men, suggest not only the recognition of persons but a hierarchy of persons, according to the relative predominance of intelligence over instinct, a hierarchy continuing the series of kinds of individuals, from buttercups to worms to seagulls, along which we have already marked off one division. And now if we consider the ethologist himself observing his subjects, whether gulls, fishes, or his fellow men, we find we have once more, without admitting any discontinuity, and granting the mirror held by gulls and sticklebacks to human nature, yet taken a new step. Tinbergen, in justifying his interest in bird watching, tries to explain it as fulfilling a kind of hunter's instinct. Doubtless it does so, but this explanation seems halting and incomplete: for it omits the thrill of personal acquaintance which he had talked of earlier, the contemplative aspect which has a share in all knowledge, but more conspicuously in disciplines concerned with the immediate knowledge of living things as living. Watching a bird-watcher watching, we are acknowledging the existence of a person in the fullest sense we know of, a person motivated not by appetite but by an intellectual passion, a passion for the understanding of other living things. Similarly, if we could watch a theoretical physicist at work we should be watching a person motivated by an intellectual passion, a passion for working with beautiful mathematical theories in the confidence that the world is as these theories dictate—that, as Norman Campbell put it, 'pure thought aiming only at the satisfaction of intellectual desires'[38] can lead us to knowledge of the external world. As the sea-gull, the stickleback or the man are driven by their individual appetites, the man who happens to be a scientist is driven also by a drive continuous with instinct yet emergent as something profoundly different from it: not an appetite consuming that which it is nourished by, but a passion which seeks intellectual satisfaction not only as what satisfies itself but with universal intent:

'The social lore which satisfies our intellectual passions is not merely desired as a source of gratification; it is listened to as a voice which commands respect. Yielding to our intellectual passions, we desire to become more satisfying to ourselves, and accept an obligation to educate ourselves by the standards which our passions have set to ourselves. In this sense these passions are public, not private: they delight in cherishing something external to us, for its own sake. Here is indeed the fundamental difference between appetites and mental interests. We must admit that both are sustained by passions and must ultimately rely on standards which we set to ourselves. For even though intellectual standards are acquired by education, while our appetitive tastes are predominantly innate, both may deviate from current custom; and even when they conform to it, they must both ultimately be accredited by ourselves. But while appetites are guided by standards of private satisfaction, a passion for mental excellence believes itself to be fulfilling universal obligations."[39]

Here then we meet full, responsible personhood. We meet responsible persons also, of course, in all our dealings with other human beings, who share with us aspirations transcending the limits of individual appetite, aspirations which may be moral as well as intellectual, practical as well as theoretical. What I have been doing here is simply to acknowledge that knowing is one of the things which we recognize responsible persons are doing, whether they happen to be knowing matter, life or other persons, electrons, rhynchodemuses, sea-gulls or themselves as doers or knowers. And this brings us full circle, or full spiral, back to our starting-point: to the unspecifiable component of our knowledge of the unspecifiable component of knowledge: to the recognition of aesthetic recognition in the knowledge of persons, individuals, and patterns.

NOTES

1. *P.K.*, p. 379.
2. C. F. A. PANTIN, 'The Recognition of Species,' *Science Progress*, 42, 1954, pp. 587–98.
3. *P.K.*, pp. 56, 62–3.
4. ibid., pp. 55–7.
5. This is a paradigm, not necessarily a piece of history. That Kekulé's reliance on structural concepts was by no means unambiguous is suggested by a paper by W. V. and K. R. Farrar, 'Faith and Doubt: The Theory of Structure in Organic Chemistry,' *Proc. Chem. Soc.*, 1959, pp. 285–90.
6. MAX BORN, *Theory and Experiment in Physics*, Cambridge, Cambridge University Press, 1943; Dover reprint, 1956, p. 12–13. Norman Campbell in *What is Science?*, Dover edition, New York, 1952, p. 156, uses the same example. Cf. *P.K.*, pp. 145–9.
7. P. DIRAC, 'Quantum Mechanics and the Aether,' *Scientific Monthly*, 58, 1954, p. 142.
8. *P.K.*, pp. 16, 166.
9. SIR HAROLD JEFFREYS, *Scientific Inference*, 2nd edition, Cambridge, Cambridge University Press, 1957.
10. ibid., p. 77.
11. R. B. BRAITHWAITE, *Scientific Explanation*, Cambridge: Cambridge University Press, 1953, pp. 163 ff.
12. G. S. FRAENKEL and D. L. GUNN, *The Orientation of Animals*, Oxford, Oxford University Press, 1940.
13. *P.K.*, pp. 348–54.
14. PROFESSOR D. LEWIS, 'The New Molecular Genetics,' Joint Biology Conference, London, 24 October 1959.
15. London, 27 September 1661. Translated in A. Wolf, *The Correspondence of Spinoza*, London, George Allen and Unwin, 1928, p. 80.
16. R. B. GOLDSCHMIDT, *Theoretical Genetics*, Berkeley, Cal.: University of California Press, 1955, p. 250 (my italics).
17. See for example N. Campbell, op. cit., pp. 53–5 or B. Russell 'On the Notion of Cause' in *Mysticism and Logic*, Pelican edition, 1953, pp. 171–96.
18. The point I am making here is related to the argument of *P.K.*, Chapter 11, especially pp. 331–2; but the role of 'causal' concepts in explanation needs further analysis along the lines suggested there.
19. WALTER M. ELSASSER, *The Physical Foundations of Biology*, New York,

The Pergamon Press, 1958. See Professor Wigner's essay on 'The Probability of the Existence of a Self-Reproducing Unit' in this volume (p. 231 ff.).
20. G. W. LEIBNIZ, *Mathematische Schriften*, ed. Gerhardt, Berlin and Halle 1849–55, vol. VI, pp. 129–35. English translation in *Gottfried Wilhelm Leibniz, Philosophical Papers and Letters*, ed. L. E. Loemker, Chicago: University of Chicago Press, 1956, Vol. I, pp. 538–43.
21. H. S. JENNINGS, *The Behavior of the Lower Organisms*, New York, Columbia University Press, 1906 (4th reprinting, 1931), p. 336.
22. *P.K.*, pp. 361–3.
22a. Dr. W. H. Thorpe, who has kindly read this paper, points out that this impression is invalidated by modern cinemaphotography of the processes of growth. Yet there does seem to be a line to be drawn somewhere!
23. Quoted by Fraenkel and Gunn, *op. cit.*, p. 4 from G. J. Romanes's *Mental Evolution in Animals*, London, Kegan Paul, 1885, p. 279. There is ample evidence of a similar subjectivism in Romanes's *Animal Intelligence*, London, Kegan Paul, 1882, as in the account of emotions in spiders, general intelligence in birds, and so on. It is interesting to note, however, that the argument in *Mental Evolution* is introduced in support of Romanes's enthusiastically Darwinian position: an instinct so often lethal must have some explanation in terms of adaptive relationships.
24. J. A. BIERENS DE HAAN, 'Animal Psychology and the Science of Animal Behaviour,' *Behaviour*, 1, 1948, 71–80.
25. N. TINBERGEN, *The Herring Gull's World*, London, Collins, 1953. A much more massive argument for the point I am trying to make here is found in W. H. Thorpe's *Instinct and Learning in Animals*, Cambridge, Cambridge University Press, 1956.
26. TINBERGEN, *op. cit.*, p. 141.
27. ibid., p. 154.
28. ibid., p. 54. Tinbergen adds: 'The ceremony is understood by other gulls.'
29. ibid., p. 60.
30. ibid., p. 137.
31. Cf. also Tinbergen's account of van Dobben's method of gull control (ibid., p. 173): 'A limited area in a colony is selected as the future site of the reduced colony. In this area all eggs are shaken, and outside the area all eggs are taken. Part of the outside gulls will start a new clutch in the same area, but part change over to the central area, where there is, *from the limited point of view of the gulls*, no egg-robbing.' (My italics.)
32. ibid., p. 51 (*my italics*).
33. ibid., p. 79.
34. ibid., p. 81.
35. ibid., p. 167. See also the reference to 'tradition' in the account of chick-stealing, p. 174.
36. ibid., p. 84–5.
37. ibid., p. 72–3.
38. Norman Campbell, op. cit., p. 156.
39. *P.K.*, Chapter 6, pp. 173–4.

18

Origin of Life on Earth and Elsewhere*

MELVIN CALVIN

Abstract

WE trace a path from the primitive molecules of the primeval earth's atmosphere *condensed from space*, through the random formation of more or less complex organic molecules, using the available energy sources of ultra-violet light, ionizing radiation or atmospheric electrical discharge, through the selective formation of complex organic molecules via autocatalysis, finally, to the information-transmitting molecule which is capable of self-reproduction and variation. In addition, somewhere, either during the course of this chemical evolution, or perhaps succeeding it, a system has been evolved in which the concentration of the reaction materials was retained in a relatively small volume of space, leading to the formation of cellular structures. Man is about to send back into space some bits of the dust from whence it originally came. It is thus not only timely but more significant than ever before to ask again the question: What are the probabilities that cellular life as we know it may exist at other sites in the universe than the surface of the earth?

We can assert with some degree of scientific confidence that cellular life as we know it on the surface of the earth does exist in some millions of other sites in the universe. We thus remove life from the limited place it occupied, as a rather special and unique event on one of the minor planets, to a state of matter widely distributed throughout the universe.

Man's adventure into space, which is about to begin, is a necessary aspect of evolution and of human evolution, in particular. It is an activity within the capability of this complex organism, man, and it must be explored as every other potentially-useful evolutionary possibility has been. The whole evolutionary process depends upon each organism developing, to the greatest extent, every potential.

Whence came life on the surface of the earth? Whether or not a complete answer to this question may be found within the context, and content, of

* The preparation of this paper was sponsored by the U.S. Atomic Energy Commission.

modern science, may be a moot question. It is our purpose to see how far we can devise an answer, and how satisfactory it may be, within that context. It becomes clear immediately that we will be dealing not only with the advances of biology that have occurred, but all of the contiguous sciences—physics, chemistry, geology, astronomy, and the like. However, our primary point of view will be that of biology and chemistry.

In trying to provide this answer we will, of necessity, have to review the accomplishments, particularly of these areas, both in their practical, concrete knowledge, as well as the impact they have, and will have, on man's view of himself and his place in the universe. At every point of our discussion, we will limit ourselves to asking questions, and providing answers, which, at some point, may be susceptible of observational or experimental test.

What is Life?

Since we have phrased the question in terms of 'the origin of life', we presumably have a clear conception of what we mean by the term 'life'. There is very little doubt that, on the ordinary level of human experience, there is no difficulty in distinguishing that which lives from that which does not. However, when we explore this notion to try to determine precisely what it is, or to be even more specific, what qualities we must devise, in order to produce something which lives from something which does not with no help from a living agency save the hand of man, the question becomes somewhat more ambiguous. For example, there are many qualities which we have no difficulty in attributing to a living organism. It is able to reproduce itself, to respond to an environmental change (i.e. 'come in out of the rain'), to transform energy into order (sunshine into a leaf, leaf into a hair), to change and remember the change, etc. While many of these individual characteristics may be duplicated in systems in space, one or more at a time, it is only when a sufficiently large collection of them appears in a single system in space that we call that system alive.

Thus the definition of life takes on the arbitrariness of the definition of any particular point on a varying continuum, and precisely where that point will fall on the line of time and evolution will depend a good deal upon who is watching the unfolding of that line.

With this last remark, we have introduced the basic notion of evolution, which, since its precise formulation exactly one hundred years ago by Darwin and Wallace, has pervaded all of science. In fact, most of what I have to say in this paper could be formulated in terms of a long extrapolation backward in time of the notions that were so ably expressed by Darwin and Wallace in 1858, but which they do not extend very far back, either in geologic time or certainly in cosmic time.[1]

As most of my readers will know, for a period of over sixty years any serious discussion of the question of the origin of life was not indulged in by scientists, particularly by experimental scientists. In fact, it was considered a disreputable kind of activity. It is interesting to examine what some of the reasons for this might have been. One certainly (and perhaps a dominant reason) was the dictum of a contemporary of Darwin's (1809–82), a chemist, Louis Pasteur (1822–95), who, in 1864, quite clearly provided an answer to the question which, for many decades previous, had been the subject of much discussion, namely, the possibility of the spontaneous generation of life on earth today. Pasteur quite clearly, and definitely, established, in his experiments of 1864, that it was impossible on the earth today, under controlled conditions, to demonstrate the appearance of living material except through the agency, or as offspring of, other living material. While Pasteur opposed the Darwinian formulation of evolution, largely on religious grounds, I suspect that he either knew consciously, or felt instinctively, that the Darwinian doctrine was conceptually in conflict with his experimental conclusion.[2] A search of the works of Darwin has revealed no mention of his opinion of the conclusions that Pasteur reached in 1864.[3]

In any case, for over sixty years thereafter, as I mentioned earlier, there appeared practically no serious discussion of the origin of life, or spontaneous generation. Between the publication in 1870 by Alexander Winchell, a professor of geology, zoology, and botany at the University of Michigan, of a book entitled, *Sketches of Creation*[4] and the statement by J. B. S. Haldane, professor of biology, in 1928,[5] there appears to be no serious attempt to answer the question of the origin of life within the context of the science of the period.

The hiatus came to an end with the recognition, by Professor Haldane, that the dictum of Pasteur was not in conflict with the backward extrapolation of the doctrine of evolution as expounded by Darwin and Wallace, if one recognizes that at the time that spontaneous generation must have occurred, according to the evolutionary extrapolation of Darwin and Wallace, there was not, by definition, any living thing on the surface of the earth. Therefore, it was possible, in the pre-biotic time, to accumulate large amounts of organic material generated by non-biological processes. This, of course, cannot take place on the surface of the earth today, since there exist everywhere on the earth's surface organisms, both micro- and macro-, which would transform any such organic material immediately it was formed, even in small amounts. Thus, the apparent conflict of concepts between the backward extrapolation of evolution and the dictum of 'no life save from life' can be, and has been, resolved.[6] Since that time, that is, 1928, it has become increasingly popular amongst scientists (experimental and otherwise) to examine the question of the origin of life from the scientific point of view. In fact, it has become so popular that within the eighteen months preceding the writing of this paper (December 1959), there were held at least two conferences in the United States and one international conference on the subject.

The Time Scale

It seems wise to have a look at the time we have in which to accomplish this total evolution of life. Here, in the first of our figures (Figure 1) we

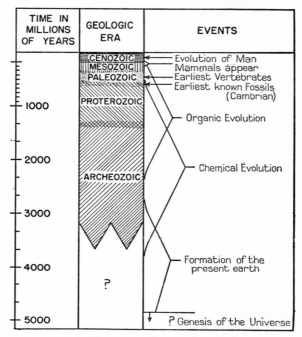

FIG. 1. Time scale for total evolution.

can see something of the order of magnitude of the time with which we have to deal related to the geologic history of the earth. The earth was formed from matter in space some four to seven billion years ago. Whether this was an aggregation of cosmic dust or a primeval explosion remains a matter of some controversy which will come up again a bit later in our discussion. However, that terrestrial history itself is only some five billion years in extent seems to be well established. You can see in Figure 1 that we have marked the evolutionary periods on the surface of the earth to correspond with the known geologic eras.

The earliest period might be spoken of as the period of evolution of the present earth. Overlapping this, and including the Archeozoic and Proterozoic geologic eras (some four to two billion years ago) is the principal period of which we will speak, namely, the period of Chemical Evolution. It was during this time that the formation of more complex organic molecules from simple ones occurred by non-biological methods which we will try to describe in a moment. Overlapping the period of Chemical Evolution we have marked the period of Organic Evolution, up to the present day. This period of Organic Evolution is the one whose later part is recorded in the form of fossils. However, for its greatest part, beginning some time in the Archeozoic period and extending through the Proterozoic period, there is no fossil record. This period of Organic Evolution of the soft-bodied living organisms, which left no fossils, constitutes by far the longer fraction of the period which we call Organic, or Biological, Evolution. Finally, at the very apex of the entire structure exists a point which we call the Evolution of Man. This constitutes only something like the last million years of geologic time, and it is clear that this represents a minute fraction of the time period with which we have to deal.

THE PRIMITIVE ATMOSPHERE

It should be noted that one of the prerequisites of all of the speculation about the origin of life is that there exists a means of gradually producing relatively complex organic substances by non-biological processes. This question is susceptible to experimental investigation, and has been investigated by a variety of experimental scientists, with positive results. It is clear that in order to test any chemical process as a possible means of generating organic material by non-biological means, we must first know what the raw materials for these chemical processes must be. This, of course, entails some knowledge of the nature of the atmosphere of the primeval earth. This, in turn, requires a concept of the mode of formation of the earth and the solar system in which it exists. There have been a wide variety of hypotheses on this point. For example, Shapley, in his recent discussion, lists some fifteen different hypotheses with regard to the origin of the earth and the solar system.[7]--- In any case, one thing is common to all of the hypotheses, namely, that the earth did have a solid crust and some kind of a gaseous atmosphere. The question that is open to some discussion is whether that crust and atmosphere were primarily oxidizing or primarily reducing in character. In the former case, the dominant partners for all the atoms are oxygen atoms, while in the latter they are hydrogen atoms.

It is clear that no matter what concept one accepts for the origin of the earth, the atmosphere itself must have been made up of relatively simple molecules such as nitrogen, ammonia, possibly carbon dioxide, methane, hydrogen, and the like. A group of these molecules is shown in the first row of Figure 2. There has been considerable discussion as to whether the oxidized molecules or the reduced molecules constituted the major portion of this atmosphere. It would appear that the present consensus favours the reduced group. In any case, experiments have been done with both types of atmosphere.

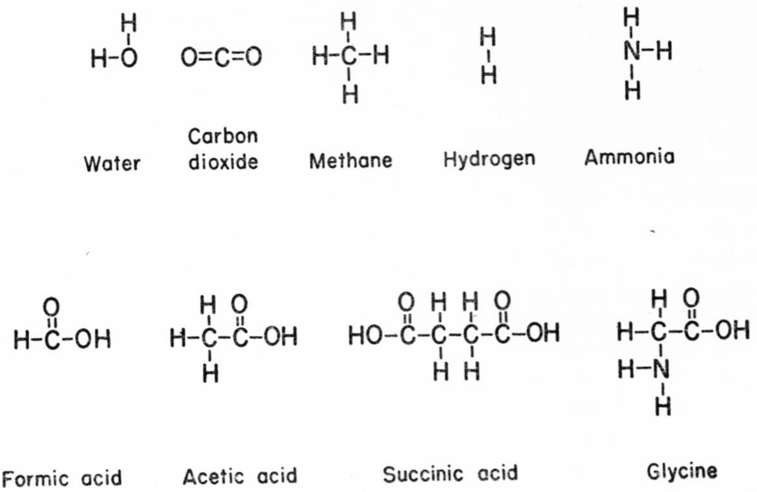

Melvin Calvin

$$H-\overset{H}{\underset{}{O}} \qquad O=C=O \qquad H-\overset{H}{\underset{H}{C}}-H \qquad \overset{H}{\underset{H}{|}} \qquad \overset{H}{\underset{H}{N}}-H$$

Water   Carbon dioxide   Methane   Hydrogen   Ammonia

$$H-\overset{O}{\underset{}{C}}-OH \qquad H-\overset{H}{\underset{H}{C}}-\overset{O}{\underset{}{C}}-OH \qquad HO-\overset{O}{\underset{}{C}}-\overset{H}{\underset{H}{C}}-\overset{H}{\underset{H}{C}}-\overset{O}{\underset{}{C}}-OH \qquad H-\overset{H}{\underset{H-N-H}{C}}-\overset{O}{\underset{}{C}}-OH$$

Formic acid   Acetic acid   Succinic acid   Glycine

FIG. 2. Primeval and primitive organic molecules.

RANDOM ORGANIC SYNTHESIS—THE BEGINNING OF CHEMICAL EVOLUTION

The agents which have been called upon to produce the initial transformations have nearly all been of the high-energy type—ultra-violet radiation, electrical discharge, radiation from radioactivity of earth-bound minerals, or radiation coming to us from outer space in the form of cosmic rays.[6] Since it appears that all of the concepts of the earth's formation involve the absence of molecular oxygen from its primeval atmosphere, the intensity of ultra-violet light which impinged on the surface of the earth in its early days was considerably greater than it is today. So, some of the earliest experiments designed to determine whether more complex organic molecules, containing carbon-carbon bonds, could be formed were done using the ultra-violet light as the source of the energy. They were described by Haldane as early as 1928 and they have been since checked in a variety of laboratories throughout the world. The primitive carbon compounds used in most of these experiments were already partly reduced, such as carbon monoxide, formic acid, or formaldehyde.[8]

In 1951, in our laboratory, experiments were again instituted to determine the usefulness of high-energy radiations, such as those which might be derived from the natural radioactive materials present in relatively high concentrations in the primitive earth, or from cosmic radiations coming in from space.[9] Here, the primitive carbon compounds were of the more oxidized type, strictly speaking, carbon dioxide. However, molecular hydrogen was also present in these experiments. The partial reduction of carbon from the completely oxidized form of carbon dioxide to a partly reduced form, such as formic acid or formaldehyde, was observed. In addition, new carbon-carbon bonds were apparently formed upon irradiation of aqueous solutions of such carbon compounds, leading to compounds such as oxalic acid and acetic acid, shown in the second row of Figure 2.

Still more recently, beginning with the premise that the primeval atmosphere was of a reducing character, experiments were undertaken which were designed to test the effectiveness of electrical discharge in the upper atmosphere (a reduced atmosphere) to create materials more closely resembling those which are presently used in biological activities.[10] Here the compounds used as starting materials were water vapour, methane, ammonia, and hydrogen. As you see, these were primarily the reduced forms of each of the elements, that is, the elements of oxygen, nitrogen, carbon, attached only to hydrogen. When an electric discharge was passed through such a mixture, one did indeed get a large variety of more complex materials, particularly those known as amino acids, of which the simplest, glycine, is shown in the figure. It is perhaps worth while to point out that having first reduced the one-carbon compound in the high-energy experiment and then formed a two-carbon compound by connecting two carbon atoms in the form of acetic acid, it was possible to demonstrate the generation of succinic acid, a four-carbon compound, resulting from the combinations of two molecules of acetic acid, by irradiation with high-energy radiations such as those which one might obtain from radioactive materials. It is interesting to note that these latter experiments, as well as the first ones of 1951, were done using a cyclotron rather than natural sources of radioactivity because of the high intensities of ionizing radiation which could be obtained.

One suggestion of a very different type was among the early ones used, since the elementary chemistry was already well known. This chemistry begins with metallic carbides,[11] which, on contact with water, will produce a variety of hydrocarbons such as methane and acetylene (the latter the familiar process used in the older miners' lamps). Some of these gases are of such a character that if they come into contact at high enough concentration with any of a variety of mineral surfaces, they will combine with each other to produce large, complex molecules, sometimes with very specific configurations.[12] The metal carbides used to start such a process generally require high temperatures for their formation, and it will therefore be necessary to suppose either that the earth began very hot or that it has had at least some places in, or on it, that were hot enough to form such carbides and later deliver them to the cooler surfaces. There appears to be no geologic evidence for the primary existence of such carbides in the deep rocks.

Very nearly all the processes which we have just described for the generation of new carbon-carbon and carbon-nitrogen bonds and the creation of more complex molecules from simpler ones are processes which depend upon the primary disruption of the simple molecule into an active fragment, followed by the random combination of those active fragments into something more complex. As the material present in the atmosphere and on the crust of the earth is gradually changed from the simple to the complex, these methods of transformation will not select between the simple and complex, and are just as prone to destroy by bond breakage the products of their initial activity as they are to create new ones. It is from here that we must now call upon some method of selective construction of molecules.

THE EVOLUTION OF CATALYSTS—ENZYMES

To do this we need only to call upon the phenomenon of autocatalysis, well known to chemists.[13] This phenomenon occurs whenever the product of any chemical transformation has the property (catalytic) of influencing the rate of its own formation. It is somewhat surprising that this phenomenon was not utilized long ago in such discussions as these, since it is, in essence, the very fact and form of the mechanism of all selective evolutionary processes, namely, the selective superiority of any form to reproduce itself that will give rise to a transformation of material into that particular form.

Thus, it is easily apparent from Figure 3 that of three possible transformations which A might undergo, namely, to B, to C, or to D, it will

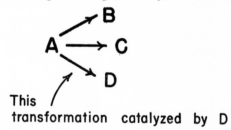

FIG. 3. Autocatalysis as the means of chemical selection.

undergo more frequently the transformation to D because D itself is a better catalyst for that transformation than it is for the others. Therefore, eventually the random processes which gave rise to A will not be quite so completely random in their further effectiveness. They will, in this case, tend to produce D rather than B or C from A. In general, the selective process of autocatalysis, in higher evolution as well, is not an 'all or none' process as we have just described it, but rather one of a matter of degree. Thus, the actual situation is more likely to be that all three substances, B, C, and D, are catalysts for their own formation, but the most efficient is the one which will eventually supersede the others.

It is of interest to examine how this sort of chemical selectivity might have functioned in the development of an extremely important biological material which is widely distributed today. This class of material is represented by that which gives the red colour to blood cells and the green

Origin of Life on Earth and Elsewhere

colour to leaves. It constitutes a very important class of organic substances known by the name of porphyrins. We have already seen how the simple precursors to porphyrins, succinic acid and glycine, might be formed (and indeed, were formed) by a random process from the primeval carbon-containing molecules. Now we have available to us a method of tracing the route by which these two precursors eventually become porphyrins in present-day living material, and some of the essential steps in that series are shown in Figure 4.[14]

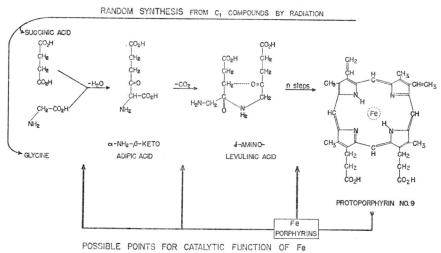

FIG. 4. Biosynthesis of porphyrin and the evolution of the catalytic functions of iron.

Underneath the centre sequence are indicated, by arrows, a number of possible points where the catalytic function of iron might play a role. For example, iron may play a part in the combination of succinic acid and glycine to start with, or in the decarboxylation of the primarily-formed keto adipic acid, which is the second compound in the centre row, or, finally, in one of the several steps which are necessary to convert the delta-aminolevulinic acid into the macrocyclic, or large ring, compound known as protoporphyrin No. 9 at the right-hand end of the figure. We already know that a number of these steps are definitely catalyzed by iron, but what is more important is that some of them are much more readily induced to go by the iron ion after it has been encased in the porphyrin No. 9, as shown by the dotted circle in the porphyrin than they are by the bare iron ion. Such an iron porphyrin is a much better catalyst for several of the steps leading to its own formation than is the bare iron ion itself. Furthermore, it is almost certainly a better catalyst for the conversion of the levulinic acid towards the porphyrin than for the competitive conversion of the glycine back towards the carbon dioxide whence it came. Thus it is clear that the route from succinic acid and glycine to the porphyrin, once the porphyrin has been formed, will be greatly facilitated by the incorporation of iron into it, thus bringing a good deal more of the succinic acid and glycine into this particular, and important, form.

The process of the development of the catalytic function of iron does not and has not stopped at this point. When the iron compound is built into a protein (a macromolecule made by the combination of many amino acids), its catalytic efficiency may be increased still more. Beyond this, the variety of chemical changes in which it may assist can be increased and the efficiency of its function diversified.

Other atoms and groups of atoms having rudimentary catalytic powers can be developed into the highly efficient and specific catalysts which we know as enzymes, in even the simplest of modern organisms, by exactly analogous processes of chemical selection by autocatalysis. And it is almost certainly no 'accident' that these enzymes are all proteins. For it is only the proteins as a class that offer the combination of simplicity of formation (peptide bond) with practically an infinite variety of chemical function (R groups)[15] (Figure 5).

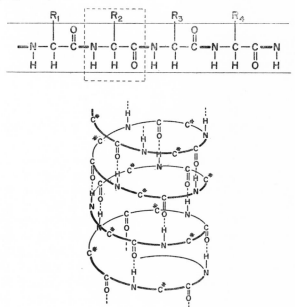

FIG. 5. Protein structure: simple structural principles and variety of chemical reactivity.

From Chaos to Order—Molecular Crystallization

So far, all of the processes of which we have spoken have been described as taking place in a rather dilute solution, with the molecules randomly arranged. The development of a complex material under such circum-

Origin of Life on Earth and Elsewhere

stances could not go very far. Two further stages of change seem to be required: (1) the ordering of the molecules in some rather specific array and (2) the concentration of the formed substances into relatively small packages. Which of these two processes took place first, or whether they were indeed successive or simultaneous developments, is hard to say. However, that both of them took place we do know, and that there exist mechanisms for each of these processes to take place we can also demonstrate in the laboratory.

One type of molecule, for example, with which we are constantly dealing, both in the laboratory and as a result of present-day biological action, is the large, flat, so-called aromatic molecule. The porphyrin that we have just mentioned constitutes one such, and there are many others. Such molecules, when they reach a certain minimal level of concentration in aqueous solution, will tend to drop out of that aqueous solution. However, they will not drop out in a random way. Because of their peculiar flat shape, they will tend to drop out in an arrangement in which the molecules are piled one top of the other, much like a pack of cards. If one throws a pack of cards in the air and allows them to fall out of the air, on to the floor, it is unlikely that they will fall standing on edge. They will, of course, practically all (if not all) land flat, one on top of the other, because of their peculiar shape. This kind of phenomenon illustrates the way in which large, flat molecules tend to come out of solution in crystals. Figure 6 shows a

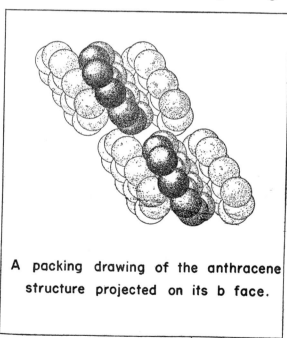

FIG. 6. Packing of flat molecules in a crystal.

diagrammatic drawing of a simple crystal of a rather simple flat molecule, and here you can see the result of the tendency of such molecules to pile up, one on top of the other.[16]

I use this particular type of example with a purpose in mind. The reason is that much of the information that present-day living organisms carry with them and can transmit to their offspring in the form of genetic material, is made up, or seems to be contained, in an ordered array of such flat molecules as this. The additional feature that one must add to this piling-up of flat molecules is that they are not independent, as the group of cards was, but that they are all tied together along an edge, as though the group of cards were tied together by a double string. The example of the flat units that we will use here will be the pairs of complementary bases as they are found in desoxynucleic acid (DNA), the material of which the chromosomes are constructed. Figure 7 shows these pairs as they occur:

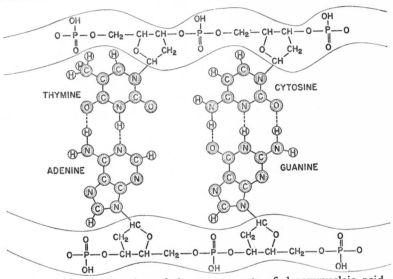

FIG. 7. Molecular drawing of the components of desoxynucleic acid—genetic material.

thymine paired with adenine and cytosine paired with guanine by virtue of their peculiar structure which places certain of their hydrogen atoms between oxygen and-or nitrogen atoms on the complementary molecule. These four molecules, and one other very closely related to thymine, are themselves formed from the same primitive precursors described in Figure 2 (succinic acid, glycine, formic acid, carbon dioxide, and ammonia). Along either edge of these flat units you can see the sugar-phosphorus chain which corresponds to the string-card analogy which I spoke of a moment ago.

Because of their peculiar aromatic type of structure these base pairs will tend to line up one above the other. This tendency will, of course, be

Origin of Life on Earth and Elsewhere

increased because they are not independent of each other but tied together by the sugar-phosphate ribbons. In actual fact, the structure of desoxynucleic acid (DNA) seems to be such a flat piling up of the nucleoside bases with the sugar-phosphate ribbons twisted in a spiral around the outside as shown in the next figure (Figure 8).

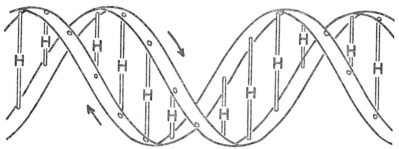

FIG. 8. Double helix model for DNA.

CONCENTRATION AND LOCALIZATION—THE FORMATION OF A CELL

The other aspect of which we spoke a moment ago, which must obtain, is the concentration into relatively small volumes in space of these organic materials, leading ultimately to the formation of cells as we now know them as the units of living matter. It is at this point that our information and our analogies are most diffuse. However, a number of physical-chemical processes have been called on to participate in the development of the living cell. First among these is the appearance of surface layers, or boundary layers, such as one sees in a soap bubble. Here we are quite familiar with mechanisms for producing a relatively stable and well defined boundary layer between two phases. Another phenomenon, which is much less familiar to us, is known by the name of coacervation and has been called upon as a primary phenomenon leading to the development of local concentration and cellular structures.[17] This phenomenon is dependent upon the ability of giant molecules in water solution to separate out from the dilute water solution into relatively more concentrated phases, or droplets, suspended in the more dilute water solution around them. Beyond this, the giant molecules tend to pack themselves in ordered arrays, provided they, themselves, have ordered structures.

As an example of this relationship between an invisible structure and the visible structures it can produce, we have selected a group of three photographs (kindly supplied through the courtesy of Professor Robley C. Williams of the Virus Laboratory). The first photograph (Figure 9) shows a number of tobacco mosaic virus (TMV) rods in their native state. These rods may, by suitable treatment, be separated into solutions of protein (approximately 95 per cent of the rods) and ribonucleic acid (RNA) (approximately 5 per cent of the rods). The RNA carries the information necessary for the reconstruction of more of the acid and the proteins by whatever cell the virus later infects, and we will discuss this genetic control in the next section.

At the moment, we are concerned with the fact that when the protein part of the TMV is put back in the proper ionic strength and pH, the molecules will reassemble themselves into rod-like structures resembling very much the original, intact particles (Figure 10). Notice, however, that in the absence of the nucleic acid the protein assembles itself into rods of random lengths and not of uniform lengths. Finally, if the nucleic acid which, as you will see later, contains the genetic information, is added back into the protein solution and then the pH and ionic strength adjusted for reconstitution, particles are obtained which are practically indistinguishable from the original TMV particles both in physical properties and biological activity (Figure 11). Thus we see how the macrostructure of visible cellular elements is implicitly contained in the microstructure of the molecules of which these elements are made.

While the knowledge of the interaction of giant, or macro, molecules, both synthetic and natural, has made great strides in the last decade or two, we are still only at the beginning of our investigation of systems of macromolecules of sufficient complexity to provide us with the kind of information we would like to have for the present purposes. However, it seems to me that already enough information is available to us to be able to say with confidence that the basic kinds of physical-chemical processes upon which we will have to draw in order to describe the evolution of the cellular structure are already at hand.

THE SELECTION OF THE PILOT DNA

That characteristic which is most frequently invoked as the prime attribute of living material is the ability to reproduce and mutate. Very frequestly in discussions of the origin of life, attempts are made to define that certain point in time before which no life existed and after which we may speak of 'living' things, as that point at which an organic unit, be it a molecule or something bigger, came into existence which could generate itself from existing precursors and which could sustain and propagate a structural change. Since we can be confident that genetic information is transmitted today in the environment of a cell by the chemical substance known as desoxynucleic acid (DNA), of which we have already spoken, it seems reasonable to seek in this structure the clues to the mutable self-reproducing molecule, or unit.[18]

It should be pointed out that while we are in the habit of thinking of the nucleoprotein molecules which constitute the chromosomes and viruses (DNA and RNA, ribonucleic acid) as containing all the information

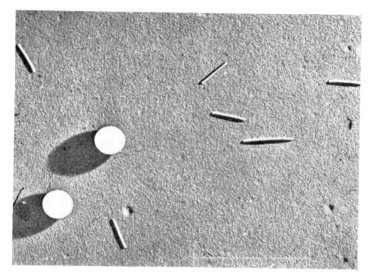

Figure 9:
Native tobacco mosaic virus.

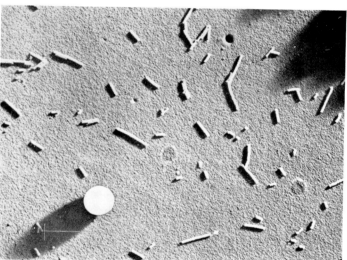

Figure 10:
Repolymerized TMV-protein.

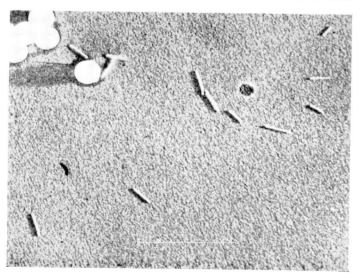

Figure 11:
Reconstructed TMV-protein and ribonucleic acid.

required to produce its entire organism, this is not strictly true. These structures may be said to contain information only in relation to, and exhibitable through, a proper environment. Thus it is possible to keep a bottle containing only virus particles (say DNA) indefinitely, just as any other organic chemical, and the question of whether it is alive or not would not arise. The moment these particles find themselves in a suitable organic medium (such as may be found in any of a variety of cell cytoplasms), this information makes itself apparent and the virus multiplies. There are thus other constituents in the modern cell which contain indispensable information but under direction of the nucleoprotein of the chromosomes.

While a chemical model[19] operating in one dimension has been described which is able to control the behaviour of a mixture of precursor units as it goes on to form a larger material, it is only in recent months that the ability to separate the controlling information-carrying units from the energy-transforming units and to demonstrate that information can indeed be carried by the desoxynucleic acid particles has been achieved.[20]

You will recall that the DNA, or chromosomal material, is made up of a linear array of only four units, represented in Figure 7, by the four bases (adenine, thymine, cytosine, and guanine) and that the controlling information about the organism (at least a modern post-Cambrian cell) is contained in some kind of linear array of these units. In the past few years it has been possible to isolate from various living organisms, particularly bacteria, a catalyst (enzyme), which, when placed into a solution containing all four of these units in an active form, that is, as their triphosphates, was able to induce their combination into some particular linear array of bases to produce a particular variety of desoxynucleic acid. Which particular DNA was formed depended entirely upon the presence of a very small amount of so-called 'starter' which had to be added to this mixture. If this starter were obtained from one type of cell, that particular type of desoxynucleic acid would be formed; from some other type of cell, another type of DNA would appear.[21]

Still more recently, it has been possible to make a synthetic desoxynucleic acid, consisting of only two of these bases, in particular the thymine and adenine.[22] This synthetic DNA presumably has no counterpart in nature today and yet when this synthetic material is given as a 'starter' to the reaction mixture, as previously described, that particular two-base desoxynucleic acid, made up of only thymine and adenine, is produced. The catalyst which is able to do this seems to be the protein-like material constructed of amino acids in the way proteins usually are. Although little is yet known about the nature of this catalyst (enzyme) it will almost certainly be related to the simpler compounds and elements whose more primitive catalytic abilities constitute the basis for its action. This is exactly analogous to the relationship between the catalytic ability of simple iron ions and that of the highly effective iron proteins, as described in Figure 4.

We have thus traced a path from the primitive molecules of the primeval earth's atmosphere condensed from space, through the random formation of more or less complex organic molecules, using the available energy sources of ultra-violet light, ionizing radiation or atmospheric electrical discharge, through the selective formation of complex organic molecules via autocatalysis, finally, to the information-transmitting molecule which is capable of self-reproduction and variation. During the course of this process we have, naturally, made use of the organic materials which have been produced in high energy form via these various energy-yielding routes. In addition, somewhere, either during the course of this Chemical Evolution or perhaps succeeding it, a system was evolved in which the concentration of the reaction materials was retained in a relatively small volume of space, leading to the formation of cellular structures.

PLOTTING THE COURSE—BIOSYNTHETIC PATHWAYS

During this entire course we have made use of the randomly-formed molecules, followed by a chemical kind of selection. The ultimate production of the information-carrying molecules depended upon the preferred presence of their constituent units, for example, nucleoside triphosphates or 'active' amino acids. It is clear that as the efficiency of transformation is increased by chemical or early biotic evolution, all of these precursors will have been used up, and a mechanism will have to be devised for the regeneration of those precursors by more specific chemical routes than those originally used. We can see these very specific biosynthetic routes in the living organism of today.

The last ten to twenty years, since the application of tracer techniques particularly, has exposed to us the wide variety of relatively complex biosynthetic sequences, an illustration of which we had in Figure 4 leading to the porphyrin, and which appear to be a sequence of reactions directed towards a particular end. The usefulness of any intermediate step does not become apparent until the final product is formed. Such sequences, on an organismic level, have led to a variety of teleological theories about the nature of evolution. However, on the molecular level it is possible to see the way in which complex, apparently directed, biosynthetic sequences arose by the operation of the ordinary laws of physics and chemistry, including the idea of autocatalysis as the basis of selection.

This was pointed out by Horowitz some years ago when he recognized that once having formed a useful material into an 'organism', which could transmit its information to its offspring, this process would continue so long as precursors were available for this 'organism' to use for its reproduction.[23] However, eventually it was clear that one or another of these precursors would become exhausted. That particular organism which could adapt itself by a random variation to make the missing precursor from molecules which still remained available to it would, of course, survive, provided the knowledge of how it was done could be transmitted; all the others would die out. Thus, we would now have lengthened the chain of synthesis by one step, but in a backward direction towards the simpler precursors. By extrapolating this back, eventually to carbon dioxide, one can get the very complex and what appear to be totally directed synthesis from the very simplest of all carbon compounds.

The Ultimate Fuel—Photosynthesis

We have remaining one additional attribute which is always associated with living material and which is very frequently called upon as a prerequisite to life, namely, the ability to use energy-yielding chemical reaactions to create order out of disorder. Ultimately, of course, the large-scale evolution of living organisms to the extent that we are now familiar with it, could not take place until the invention of photosynthesis, that is, the coupling of the ability of certain molecules to absorb solar energy, to the ability of certain other molecules to use this energy for the synthesis of the necessary structures. Here it is almost certain that the ability to synthesize had long been evolved before the ability to couple the absorbed solar energy to those synthetic reactions was discovered. The use of porphyrins by nonphotosynthetic organisms is widespread. Almost certainly random variations in their structure led to the biological discovery of the structure of chlorophyll, and its proved efficiency in transmitting energy for biosynthesis led to its evolutionary selection as the pigment of photosynthesis.[24]

Strictly speaking, the primitive synthesis of which we spoke, making use of ultra-violet light or ionizing radiation, is a form of photosynthesis, and in all probability there existed a parallel evolutionary development of this kind of energy conversion process. In fact, modern work by physical chemists[25] on the effect of the far ultra-violet light on some of the simple molecules we spoke of earlier, as constituting the primitive atmosphere of the earth, has demonstrated experimentally the feasibility of the idea of the conversion of water (H_2O) into hydrogen (H_2) and oxygen (O). It has been possible to demonstrate the conversion of carbon dioxide (CO_2) into carbon monoxide (CO) and oxygen (O) using sunlight of such a high energy that very little of it penetrates down very far into the present earth's atmosphere.[26]

Whether the conjunction of the use of visible light-absorbing substances, such as the porphyrins, with the biosynthetic demands of the more highly evolved chemical systems took place before or after the appearance of what we would today call living organisms, matters little for the purposes of our present discussion, important though the question may be. It seems quite clear, however, that these two parallel lines of development did meet, as mentioned earlier, giving rise to the very efficient energy conversion processes resembling those which we know today. It is not unlikely that the final step in the development of modern photosynthesis, namely, the evolution of oxygen, did not take place until relatively late in the sequence of events. For example, we do have organisms today which are capable of using solar energy, via the agency of porphyrin-type molecules, but using other methods of taking care of the oxygen by combination with suitable reducing agents such as hydrogen. These appear in the form of photo-reducing organisms such as the photosynthetic bacteria. They dispose of the oxidizing fragment of the water molecule, made by the absorption of light, by combining it with whatever reducing materials may have been present in the primitive atmosphere. It is not until the higher green plants appear that we find the ability of disposing of this oxygen back into the atmosphere as molecular oxygen.

The Fossil Record—Man's Place

With this biological discovery the stage was set for the enormous development of living organisms on the surface of the earth as we know them today. From here on, the fossil record is quite complete and there is little point for us to pursue in detail the ascent and divergence of life, leading ultimately to mankind in the last million years.[27]

It is perhaps worth while at this point to try and assess the amount of time which may be allotted to each of the major kinds of operations, or sequences of events, which we have outlined as leading ultimately to cellular life as we know it on the earth today, and to man. Most of you are undoubtedly aware of the fact that from the very beginning the fossil record contains evidence of very nearly all of the major sub-kingdoms, or phyla, of life today. This fossil record, which is some 500 to 1,000 million years old, thus indicates that by the time life was sufficiently well developed to leave a fossil record it had already manifested itself in nearly all of the major types of forms which we now recognize. This time period constitutes something less than one-quarter of the entire habitable life span of the earth. Thus we have some two or three billions of years during which we can pursue the process of Chemical Evolution, overlapping with that part of the evolution of cellular life (biotic evolution) which was unable to leave a record of itself in the rocks. This is an extremely long period of time, which gives us ample opportunity for the enormous numbers of trials and errors which would be required to develop all the possible molecular processes and combinations which must have been tried.

Undoubtedly, many different information-carrying molecular species had a birth, a life span and a death, much as we now see in the fossil record for the higher forms. One (DNA) or only a few, of these closely related information-carrying molecules, or molecular species, eventually superseded all the rest, because of the particular structural, chemical and dynamic properties of this arrangement of atoms, i.e. stability, template of complementarity quality, mutability and others not yet defined. It was from this (these) that the present-day organisms have developed, thus providing a basic similarity in all living processes as we know them today.

The determinism in the arrangement of a system increases with the number of trials (events) that can occur in reaching it. On the molecular level, where the number of changes occurring per second is high, predictability with regard to what will happen in a given situation to a group of molecules amounts to certainty. For example, pressure maintained by bombardment of molecules on the walls of a vessel containing a certain number of gas molecules at a specified temperature is quite predictable. While we cannot specify the particular molecules which will strike the wall at any time, we can be quite sure that in a given period of time a certain number will do so.

At the other extreme, the segregation and recombination of genes which may take place in the formation of a new individual by (geologically) modern genetic mechanisms appears as a completely undetermined, or random, choice, since only a single event is involved. However, among a group of organisms of a given species there will be a predictable distribution of properties at a given time under a specified set of conditions.

On the other hand, coming up from molecules and molecular aggregates, we will reach a stage, probably after the invention of cellular heredity based on nucleic acid, when the number of events (rate multiplied by time available) will not be large enough to ensure that all arrangements possible will have been tried, and an indeterminism with respect to those that have appeared will ensue.

Therefore, while an indeterminism exists with respect to the character of any individual living thing and a limited indeterminism exists with respect to species, the time element is so great and the amount of genetic material which has been cycled through the sequence of birth, growth, development, and death is so enormous,[28] that the certainty of the occurrence of cellular life as we know it on earth today seems assured, given the initial starting conditions.

A most convincing demonstration that such a sequence of events, leading from non-living matter to life, could, and probably did, take place, would be an experiment in which a system of organic material, called alive by most biologists, is produced through the agency of no other life save the hand of man.[29] And already today there is serious discussion as to whether some of the experiments performed in the last year might not fulfill these conditions.

It thus appears that man is a rather late and highly developed (perhaps the most highly developed) form of that organization of matter which we call living, on the surface of the earth, and which is the result of the peculiar and special environmental situation provided on the surface of the earth since its formation some five billion years ago. We have known for some time now that the earth is the number three planet in orbit around a rather ordinary star on the edge of one of the minor galaxies of the universe. Thus presuming life to be a unique occurrence limited to the surface of this one rather trivial (in terms of mass, energy, position, etc.) planet in the universe, man, though an impressive representative of the state of matter called living, is not viewed as a major cosmic force.

Life on Other Planets?—A Cosmic Influence

Now man is about to send back into space some bits of the dust from whence he originally came. This return will be in the form of machines, and eventually of man himself. It is thus not only timely but more significant than ever before to ask again the question: What are the probabilities that cellular life as we know it may exist at other sites in the universe than the surface of the earth? In view of the chemistry of carbon (and a few of its near neighbours) and the consequences this has given rise to in the environment to be found on the surface of the earth, all that is required to answer, or to provide some kind of answer to this question, would be an estimate of the number of other sites, or planets, in the universe which might have the environmental conditions within the range to support cellular life as we know it on the surface of the earth.

Here we follow Shapley (as well as others)[30], who begins his calculation with an estimate of the minimum number of stars which might be in the universe. His minimum calculation, based upon the best telescopes we have today, is of the order of 10^{20}, that is, one with twenty zeros after it. The next step is to determine what fraction of these stars will have a planetary system. To make this estimate one must have some concept of how our planetary system was generated. Here there are many theories and Shapley lists fourteen or fifteen of them. However, there are a number of conditions which must be fulfilled and these, taken together with the fact that the universe has been expanding, and some five billion years ago when the planetary systems were formed, the universe must have been much more crowded than it is today, gives the figure one in a thousand as a very conservative estimate of the number of stars which have a planetary system.

The next question which Shapley attempts to answer is what fraction of these planetary systems will contain a planet at approximately the correct distance from its star such that it will have a temperature variation compatible with the cellular life that we are seeking to determine. Here, again, we arrive at the very small, and what we consider conservative, figure of only one of these planetary systems in a thousand as containing a planet at approximately the right distance from its sun. The next circumstance which must be fulfilled is that of size. Again, the planet must be of the correct size to retain an atmosphere and not be too large. Here Shapley gives the figure of one in a thousand as a conservatively small fraction.

Finally, of all of those planets which are the correct size, which of them will have the proper atmosphere containing carbon, hydrogen, nitrogen, and oxygen, which will be required to give rise to cellular life as we know it on earth? The answer is, again, the very conservative number of only one in a thousand.

Thus, we have four factors of one-thousandth by which we must multiply our figure of 10^{20} (the minimum number of stars which might be in the universe). This leads to the number 10^8, that is, one hundred million, as the number of habitable planets to be found in the universe. Remember that this calculation is limited to those planets which will have conditions within the range compatible with cellular life as we know it on earth based on carbon. This does not include such systems, which conceivably we can imagine, based on other elements, such as silicon, or nitrogen, or perhaps even anti-matter. Such worlds and such systems may very well exist. However, these are not included as possibilities in this calculation. We have already cut down the limits by our four factors of one-thousandth, leading to something of the order of 100,000,000 planets which can support cellular life as we know it on the surface of the earth.

Since the time element seems to be about the same for all parts of the universe, namely, something greater than five billion years, it would appear that we can be reasonably certain that these hundred million other planets will indeed have cellular life on them. Since in the course of the chemical and biotic evolution, the appearance of man on the surface of the earth has occupied only a very small fragment of time, namely, one million years of the five billion, it is clear that we may expect to find cellular life, and perhaps precellular life and posthuman life, in many of these other planets. The one-million-year period, given to the evolution of man on earth, constitutes an extremely small element of this long period of time and an uncertainty of this order could, and almost certainly does, exist in the status of the evolutionary sequence elsewhere in the universe.[31]

We can thus assert with some degree of scientific confidence that cellular life as we know it on the surface of the earth does exist in some millions of other sites in the universe. This does not deny the possibility of the existence of still other forms of matter which might be called living which are foreign to our present experience.

By answering our second question in this way, we have now removed life from the limited place it occupied a moment ago, as a rather special and unique event on one of the minor planets around an ordinary sun at the edge of one of the minor galaxies in the universe, to a state of matter widely distributed throughout the universe. This change induces us to reexamine the status of life on the surface of the earth. In doing so we find that life on the surface of the earth is not a passive, existing thing, but actually changes and forms the environment in which it grows. The surface of the earth has indeed been completely transformed in its character by the development of the state of organization of matter which we call life.

Furthermore, it is undergoing another, and perhaps a more profound transformation, by one representative, or manifestation, of that organization of matter, mankind.

MAN IN SPACE—THE NEXT STEP

Now that man has the capability of taking his machines and himself off of the surface of the earth and of beginning to explore outer space, there is no reason to suppose that life, and man as its representative, will not transform any planet, or any other astral body upon which he lands, in the same way, and perhaps even in a more profound way, than he has transformed the surface of the earth. For example, it might suit him in the future to change the course of the orbit of the moon, and it seems within the realm of possibility that he should be able to do so. When we realize that other organisms may be doing similar things at some millions of other regions in the universe, we see that life itself becomes a cosmic influence of significant proportions, and man, as one representative of that state or organization of matter, becomes a specific cosmic influence himself. Thus we have come to a complete inversion of our view of the place of life and of man in the universe from a trivial to a major cosmic influence. And we have come to this view entirely upon the basis of experimental and observational science and scientific probability.

Man's adventure into space, which is about to begin, is not merely a flexing of muscles—a demonstration of strength. It is a necessary aspect of evolution and of human evolution in particular. It is an activity within the capability of this complex organism, man, and it must be explored as every other potentially-useful evolutionary possibility has been. The whole evolutionary process depends upon each organism developing to the greatest extent every potential.

NOTES

1. C. DARWIN and A. R. WALLACE, *Journ. Proc. Linnean Soc. London (Zoology)*, *3*, 1858, pp. 35–62.
2. PAUL SEARS, *Charles Darwin*, New York, Charles Scribner's Sons, 1950.
3. From *Notes and Records of the Royal Society of London*, 4, No. 1, 1959. Some Unpublished Letters of Charles Darwin—p. 13 of the journal: One of the letters (41) so far as known is the last which he dictated and signed before he died. It is all the more interesting from its contents which show that Darwin believed that the origin of life would be found to be subject to natural laws.

   (Charles Darwin to George Charles Wallich. March 28, 1882.)

   'My dear sir:
   'You expressed quite correctly my views where you say that I had intentionally left the question of the Origin of Life uncanvassed as being altogether *ultra vires* in the present state of our knowledge, and

that I dealt only with the manner of succession. I have met with no evidence that seems in the least trustworthy in favour of so-called spontaneous generation. I believe that I have somewhere (but cannot find the passage (see Appendix VIII below)) that the principle of continuity renders it probable that the principle of life will hereafter be shown to be a part, or a consequence, of some general law; but this is only conjecture and not science. I know nothing about the Protista, and shall be very glad to read your lecture when it is published, if you will be so kind as to send me a copy. I remain, dear sir,
    'Yours faithfully,
        Cf. *Life and Letters*, 3, (1871), p 18: 'C. Darwin.'

'It is often said that all the conditions for the first production of a living organism are now present, which could ever have been present. But if (and oh! what a big if!) we could conceive in some warm little pond, with all sorts of ammonia and phosphoric salts, light, heat, electricity, etc. present, that a protein compound was chemically formed ready to undergo still more complex changes, at the present day such matter would be instantly devoured, or absorbed, which would not have been the case before living creatures were formed.'

4. ALEXANDER WINCHELL, *Sketches of Creation*, New York, Harper and Brothers, 1870.
5. Collection of essays on *The Origin of Life. New Biology*, No. 16, Harmondsworth, Penguin Books, Ltd., 1954.
6. A. I. OPARIN, *The Origin of Life*, 3rd English edition, Edinburgh, Oliver and Boyd, 1957. See also, *Molecular Structure and Biological Specificity*, ed. by L. Pauling and H. A. Itano. Publication No. 2 of American Institute of Biological Sciences, Washington, D.C., 1957.
7. (a) HARLOW SHAPLEY, *Of Stars and Men*, Boston Mass., Beacon Press, 1958. (b) FRED HOYLE, *The Nature of the Universe*, Oxford, Basil Blackwell, 1955, first published in England in 1950.
8. See note 5 above.
9. W. M. GARRISON, D. C. MORRISON, J. G. HAMILTON, A. A. BENSON, and M. CALVIN, *Science*, *114*, 1951, p. 416.
10. S. L. MILLER, *J. Am. Chem. Soc.*, 77, 1955, p. 2351. Also, S. L MILLER and H. C. UREY, *Science*, *130*, 1959, p. 245.
11. See note 5 above.
12. G. NATTA, 'Stereospecific Catalysis and Stereoisomeric Polymers,' Speech at opening conference, XVI Cong. Pure and Appl. Chem., Paris, France, July 1957.
13. M. CALVIN, *American Scientist*, *44*, 1956, p. 248.
14. D. SHEMIN, *Harvey Lectures*, *50*, 1954–5, p. 258.
15. F. H. C. CRICK, *Soc. Exptl. Biol. (G.B.)*, *12*, 1958, p. 138.
16. M. CALVIN, Brookhaven National Laboratory Biology Conference No. 11. *Report No. BNL-512 (C-28)*, 1958, p. 160.
17. See note 5 above.
18. J. D. WATSON and F. H. C. CRICK, *Nature*, *171*, 1953, p. 964.
19. L. S. PENROSE and R. PENROSE, *Nature*, *179*, 1957 p. 1183.
20. M. J. BESSMAN, I. R. LEHMAN, J. ADLER, S. B. ZIMMERMAN, E. S. SIMMS, and A. KORNBERG, *Proc. Nat. Acad. Sci.*, *44*, 1958, p. 633; J. ADLER, I. R. LEHMAN, M. J. BESSMAN, E. S. SIMMS, and A. KORNBERG, *Proc. Nat. Acad. Sci.*, *44*, 1958, p. 641; A. KORNBERG, *Abstr. 134th Nat. Mtg. Amer. Chem. Soc.*, Chicago, Illinois, Abstr. 21C; M. GRUNBERG-MANAGO, P. J. ORTIZ and S. OCHOA, *Science*, *122*, 1955, p. 907.

21. S. R. KORNBERG, I. R. LEHMAN, M. J. BESSMAN, E. S. SIMMS, and A. KORNBERG, *J. Biol. Chem.*, *233*, 1958, p. 159; I. R. LEHMAN, M. J. BESSMAN, E. S. SIMMS, and A. KORNBERG, *J. Biol. Chem.*, *233*, 1958, p. 136; M. J. BESSMAN, I. R. LEHMAN, E. S. SIMMS and A. KORNBERG, *J. Biol. Chem.*, *233*, 1958, p. 171.
22. H. K. SCHACHMAN, private communication.
23. N. HOROWITZ, *Proc. Nat. Acad. Sci.*, *31*, 1945, p. 153.
24. M. CALVIN, *Science*, *130*, 1959, p. 1170.
25. R. G. W. NORRISH, Liversidge Lecture, *Proc. Chem. Soc. London*, September, 1958, p. 247.
26. BRUCE MAHAN, private communication.
27. G. G. SIMPSON, *Meaning of Evolution*, New Haven, Conn. Yale University Press, 1950; also, T. DOBZHANSKY, *Evolution, Genetics and Man*, New York, John Wiley and Sons, Inc., 1955.
28. C. P. SWANSON, *Cytology and Genetics*, Englewood Cliff, N.J., Prentice-Hall, Inc., 1957, Chapter 12.
29. H. REICHENBACH, *The Rise of Scientific Philosophy*, Berkeley, Cal., University of California Press, 1951, p. 202.
30. See note 7a and b above.
31. M. CALVIN, *Evolution*, *13*, 1959, p. 362.

General references on the subject of Evolution:
1. *Evolution. Symp. VIII, Soc. Exptl. Biol. (G.B.)*, Cambridge University Press, 1953; particularly article by J. W. S. PRINGLE on 'The Origin of Life.'
2. JULIAN HUXLEY, *Evolution in Action*, New York, Harper Brothers, 1955.
3. H. F. BLUM, *Time's Arrow and Evolution*, 2nd edition, Princeton N.J., Princeton University Press, 1953.

19

The Probability of the Existence of a Self-Reproducing Unit

EUGENE P. WIGNER

General Remarks

In his 'Analytical Study' of life and the multiplication of organisms,[1] Elsasser analyses the way in which the information is stored in the germ-cells which enables these germ-cells to develop into organisms similar to the parent—similar also in their ability to produce in their turn, germ-cells containing the same type of information. Although no clear-cut proof is presented, a good deal of weighty evidence is adduced[2] to show 'that the structure of a butterfly, a snake, a tree, or a bird cannot be deduced mathematically from some relatively compact body of basic data stored in the chromosomes'; the 'maintenance of information is . . . not adequately described in terms of the mechanistic approximation'. The present writer has also been baffled by the miracle that there are organisms —that is, from the point of view of the physical scientist, structures— which, if brought into contact with certain nutrient materials, multiply, that is, produce further structures identical with themselves. He felt that it is, according to the known laws of physics, infinitely unlikely that structures of this nature exist and the present article is a report on the considerations and calculations which he undertook in this connection.[3] Actually, the point of view is somewhat different from Elsasser's: Elsasser considers the way in which the information necessary to develop the adult specimen is stored in the germ-cells and shows that the germ-cells do not have properties which the physicist would expect to be suitable for storing large amounts of information. We shall be concerned, on the other hand, with what appears to be a miracle from the point of view of the physicist: that there are structures which produce further identical structures.[4]

Elsasser's book does not spell out very explicitly a proposition for the resolution of the problem to which he points. He postulates, on the one hand, further laws of nature not contained in the laws of physics and quantum mechanics. He calls these laws *biotonic*. On the other hand, he maintains that the validity of the laws of quantum mechanics is not impaired by the new laws. Inasmuch as quantum mechanics claims to apply to all situations, and to provide all meaningful predictions, this seems to be a contradiction. The only resolution of this contradiction which appears consistent with the ideas expressed by Elsasser is that the biological units are so complicated that it is *in principle* impossible to calculate their behaviour on the basis of the laws of quantum mechanics. Elsasser concludes that a computing machine which could store all the information contained in a germ-cell would be inconceivably large. Even this interpretation is questionable: it is quite possible that one can deduce, by abstract reasoning, consequences of the quantum mechanical equations which either reproduce, or contradict, important properties of living organisms. The present study will make an attempt in this direction. After all, we can explain, by means of quantum mechanical theory, at least a large number of the properties of solids and liquids and the accurate description of these does contain as much information as can be stored in a germ-cell. Alternately, one could derive, again by abstract reasoning, intermediate theorems which would greatly simplify the task of the computing machines. For all these reasons, it does not appear very likely that the aforementioned contradiction can be resolved. It is more likely that the present laws and concepts of quantum mechanics will have to undergo modifications before they can be applied to the problems of life.

For reasons which are not quite clear, the phenomenon of consciousness has become tabu in scientific discussions. Nevertheless, as one can see, for instance, from Neumann's brilliant discussion of the process of quantum mechanical measurement,[5] even the laws of quantum mechanics itself cannot be formulated, with all their implications, without recourse to the concept of consciousness.[6] It is very likely that those who deny the reality of consciousness only mean that the external world can be completely described without reference to the consciousness of others, that is, that the motion of matter (in the broadest sense of this word) is not influenced by consciousness, even though consciousness is obviously influenced by the motion of matter. According to the view attributed in the preceding paragraph to Elsasser, this view is neither correct, nor false, but meaningless because it is, even in principle, impossible to describe the motion of all matter by the laws of physics. In particular, it should be impossible to describe living matter in terms of the laws of physics. It seems more likely, however, that this view is incorrect and that living matter is actually influenced by what it clearly influences: consciousness. The description of this phenomenon clearly needs incorporation into our laws of nature of concepts which are foreign to the present laws of physics. Perhaps the relation of consciousness to matter is not too dissimilar to the relation of light to matter, as it was known in the last century: matter clearly influenced the motion of light but no phenomenon such as the Compton effect was

The Probability of the Existence of a Self-Reproducing Unit

known at that time which would have shown that light can directly influence the motion of matter. Nevertheless, the 'reality' of light was never doubted.

CALCULATION OF THE PROBABILITY THAT THERE BE REPRODUCING STATES

The preceding, very general and speculative considerations will be supported below by a definite argument. As will be discussed in the last section, the argument to be presented is not truly conclusive. Nevertheless, it is at least indicative. It purports to show that, according to standard quantum mechanical theory, the probability is zero for the existence of self-reproducing states. The discussion of this result and of its implications will be reserved for the last section; the present section will contain only the derivation of the result. The derivation is not a rigorous one: it will be based on an assumption which is analogous to our belief that in no system of any complexity is there any 'accidental degeneracy'.[7] It is even more closely similar to the assumption on the basis of which the second law of thermodynamics was derived.[8] The assumption is that the Hamiltonian which governs the behaviour of a complicated system is a random symmetric matrix, with no particular properties except for its symmetric nature. It is by assuming this property for the Hamiltonian, when written in the co-ordinate system in which the observables are diagonal, that Neumann proved the second law of thermodynamics to be a consequence of quantum mechanical theory. A second, probably less important, assumption will be introduced later.

The calculation will be carried out in two steps. First, it will be assumed that 'the living state' is completely given in the quantum mechanical sense: it has one definite state vector with the components v_k. Clearly, there must exist at least one state of the nutrient which permits the organism to multiply. The state vector of this state will be denoted by w. There should be, as a matter of fact, many states of the nutrient on which the organism can feed but a contradiction will be obtained already by assuming a single such state. Before multiplication, the state vector of the system, organism + nutrient, is

$$\Phi = v \times w \tag{1}$$

the cross denoting the direct (Kronecker) product. When multiplication has taken place, the state vector will have to have the form

$$\Psi = v \times v \times r, \tag{2}$$

that is, two organisms, each with the state vector v, will be present; the vector r describes both the rest of the system which is the rejected part of the nutrient and also the position, etc., co-ordinates of the two organisms. Introducing a co-ordinate system in Hilbert space which corresponds to the decomposition (2), this reads

$$\Psi_{\kappa\lambda\mu} = v_\kappa v_\lambda r_\mu. \tag{3}$$

Q

The first index (κ) substitutes for the variables which describe the part of the system contained in the 'parent', the second index (λ) substitutes for the variables which describe the 'child', the last index (μ) substitutes for the variables in terms of which the rejected part of the nutrient is described. In the same co-ordinate system, (1) reads in terms of its components

$$\Phi_{\kappa\lambda\mu} = v_\kappa w_{\lambda\mu} \tag{4}$$

and we have a double index for the specification of the state of the nutrient.

The two assumptions which were mentioned before will now be introduced in precise language. The second and less important of these is the replacement of Hilbert space by a finite dimensional space. In particular, the space of the organism shall have N dimensions, the space which describes the rejected part of the nutrient shall have R dimensions. Then κ and λ can assume N values each, μ can assume R values; the state of the organism is a vector in N-dimensional (rather than infinite dimensional) space. Since no assumptions will be made concerning the magnitude of N and R, both of which surely must be assumed to be very large, this assumption appears harmless enough. It is made to make the mathematical analysis easy and can be justified since, as the total energy available is finite, both parts of the system are restricted to a finite number of states. The state of the nutrient is a vector in NR dimensional space; i.e. it is a much more specialized state than the state of the organism. This is surprising at first but must be true since the life and multiplication of the organism is connected with an increase in entropy, i.e. the final state is less specialized than the original state.

The more relevant and more questionable assumption is that the 'collision matrix' which gives the final state resulting from the interaction of the organism and the nutrient, and which will be denoted by S, has no particular properties but is a random matrix. Since it transforms the Φ of (4) into the Ψ of (3), we have

$$v_\kappa v_\lambda r_\mu = \sum_{\kappa'\lambda'\mu'} S_{\kappa\lambda\mu;\kappa'\lambda'\mu'} v_{\kappa'} w_{\lambda'\mu'} \tag{5}$$

S, so to say, embodies the laws of interaction between any state of the material which makes up the organism and any state of the material which makes up the nutrient. 'Any state' must, however, be interpreted in the light of the preceding remark; it is any of a finite number of states. S is completely determined by the laws of quantum mechanics; it is the quantity which, according to Elsasser, actually cannot be calculated. What we shall ask is, then: given an S, is it in general possible to find N numbers v_κ, which, together with suitably chosen R numbers r_μ and NR numbers $w_{\lambda\mu}$, satisfies (5). We shall decide this question simply by counting the number of equations and the number of unknowns. We shall find that the former number is greater.

Since (5) must be valid for any κ, λ and μ, we have actually N^2R complex, or $2N^2R$ real equations. There are several identities between these but since N^2 is a tremendously large number, this fact will be disregarded.

The Probability of the Existence of a Self-Reproducing Unit

There are N unknown v-components, r has R unknown components and w has NR unknown components. Altogether, there are $N + R + NR$ complex or twice as many real unknowns—very much fewer than equations—so that 'it would be a miracle' if (5) could be satisfied.

Evidently, the preceding calculation cannot be taken too seriously because surely an organism is not completely determined in the quantum mechanical sense. There must be many states v all of which represent a living organism. Their number will be denoted by n. This number is, however, much smaller than the number N of states which the matter consituting the organism can assume: evidently most of the states of this matter are not living. Hence $n \ll N$. Let us denote the n vectors which represent living organisms by v^k, the index k running from 1 to n. Then every linear combination of the v^k will also represent a living state. As a result, n initial states will have to be considered

$$\Phi^{(j)} = v^{(j)} \times w \quad \text{or} \quad \Phi^{(j)}_{\kappa\lambda\mu} = v^{(j)}_\kappa w_{\lambda\mu}. \tag{6}$$

The final state, obtained from the interaction of the state $v^{(j)}$ and the nutrient (for which we continue to postulate only a single state w), will be denoted by $\Psi^{(j)}$. As far as the dependence of $\Psi^{(j)}$ on the first two indices is concerned, this may be an arbitrary linear combination of the n^2 vectors $v^{(k)} \times v^{(l)}$. Hence

$$\Psi^{(j)}_{\kappa\lambda\mu} = \sum_{kl} u^{jkl}_\mu v^{(k)}_\kappa v^{(l)}_\lambda \tag{7}$$

and we have instead of (5)

$$\sum_{kl} u^{jkl}_\mu v^{(k)}_\kappa v^{(l)}_\lambda = \sum_{\kappa'\lambda'\mu'} S_{\kappa\lambda\mu;\kappa'\lambda'\mu'} v^{(j)}_{\kappa'} w_{\lambda'\mu'}. \tag{8}$$

This equation must be valid for every j and, as before, for every κ, λ and μ. Altogether, we have nN^2R complex or $2nN^2R$ real equations.

The unknowns are the v, the w and the u. There are nN quantities v and, as before, NR quantities w. The number of constants u is, on the other hand, n^3R because j, k, l can assume n values each and μ can assume R values. Disregarding the possibility that neither the unknowns nor the variables may be independent of each other, the number of unknowns will be equal to the number of variables if

$$nN^2R = nN + NR + n^3R \tag{9}$$

Because of $n \ll N$, the left side is still very much larger than the right side and the equation is not satisfied. Even if one allows for the possibility that *one* of the products be arbitrary, (i.e., that $v^{(l)}$ in (8) be any state, living or not living), the number of unknown u will increase only to n^2NR and the right side will still be much smaller than the left. We arrive at the result that, if the interaction S is not 'tailored' so as to permit reproduction, it is infinitely unlikely that there be *any* state of the nutrient which would permit the multiplication of any set of states which is much smaller than all the possible states of the system.

As was mentioned before, the preceding calculation disregards the fact that the equations (8) are not independent of each other and also the fact

E. P. Wigner

that the unknowns cannot be chosen freely. Both relations are consequences of the unitary nature of S. A more detailed calculation shows, however, that the relations between the equations are just about equal to the number of relations between the unknowns and, as long as $n \ll N$, neither of them affects (9) appreciably. The conclusion of the preceding paragraph stands, therefore, in spite of these relations.

Limitations of the Preceding Calculation

Even the preceding calculation, which assumes that many quantum mechanical states represent life, is far from realistic. The difficulty is that it stipulates that at least one organism *surely* survives the interaction with the nutrient. There is no clear reason to believe this. A realistic model would permit, rather, any final state, but would demand that the sum of the probabilities of the states with two living organisms be well in excess of $\frac{1}{2}$. This would lead, instead of to (8), to certain inequalities the mathematical discussion of which is much more difficult than the discussion of the equalities (8) and has not been concluded and will not be reproduced here. It remains noteworthy that the chances are nil for the existence of a set of 'living' states for which one can find a nutrient of such nature that the inter-action *always* leads to muliplication.

The preceding result seems to be in conflict with von Neumann's well known construction of self-duplicating machines.[9] If one tries to confront the evidence of the preceding section with von Neumann's explicit construction, one finds that such a confrontation is not possible because the model used by von Neumann (based on Turing's universal automaton), can assume only a discrete set of states whereas all our variables (v. w) are continuous. This permits the postulation of an ideal behaviour of the system and the 'tailoring' of what substitutes for equations of motion in such a way that they permit reproduction. The question which is in the foreground of the present discussion is whether the real equations of motion can be expected to give reproduction. The difference between a truly macroscopic 'hard' system, for which one can assume with a little goodwill any law of motion, and the 'soft' systems which really undergo multiplication, has been stressed already by Elsasser.[10] Actually, the inapplicability of his model to biological considerations was also recognized by von Neumann.

The second piece of conflicting information is the model which Crick and Watson suggested for reproduction and which proposes a definite mechanism for the transfer of the properties to the progeny.[11] This model is also based on classical rather than quantum concepts. It is indeed an ingenious and realistic-looking model which suggests the view that it may have been difficult to find a system with equations of motion which permit reproduction, but, in spite of the adverse odds, the difficult feat has in fact been accomplished. It is not intended to contradict this view absolutely. It is necessary to point out, nevertheless, that the details of the functioning of the model do not appear to have been worked out completely. Similarly, the reliability of the model, that is the probability of its malfunctioning, has not been evaluated and compared with experience. One may incline, therefore, to the view, implicit also in Elsasser's ideas, that the type of reproduction for which the model of Crick and Watson seems to apply can, as can all similar processes, be described by the known laws of nature, but only approximately. However, the apparently virtually absolute reliability of the functioning of the model is the consequence of a biotonic law.

The writer does not wish to close this article without admitting that his firm conviction of the existence of biotonic laws stems from the overwhelming phenomenon of consciousness. As to the arguments presented here, they are suggestive but not conclusive. The possibility that we overlook the influence of biotonic phenomena, as one immersed in the study of the laws of macroscopic mechanics could have overlooked the influence of light on his macroscopic bodies, is real. This does not, however, render the arguments here presented conclusive. because, in their present form, they are based on the assumption that the laws of reproduction are absolute. This may be just as little true, or may be just as misleading, as was Leibniz's conclusion of the incorrectness of atomic theory which he inferred from the impossibility of finding two identical blades of grass.[12]

NOTES

1. WALTER M. ELSASSER, *The Physical Foundations of Biology*. London, Pergamon Press, 1958.
2. Since the whole book (cf. Ref. 1) is built around this theme, it is not possible to point to definite passages which contain all the evidence presented. Nevertheless, pages 124–32 are perhaps most characteristic of the trend of thought
3. The results of these considerations were mentioned already in the author's article 'The Unreasonable Effectiveness of Mathematics in the Natural Sciences', *Communications in Pure and Applied Mathematics*, 13, 1960, p. 1.
4. M. POLANYI's review of G. Himmelfarb's *Darwin and the Darwinian Revolution*, *The New Leader*, 31 August 1959, p. 24, expresses similar doubts concerning the possibility of explaining the phenomenon of life on mechanistic grounds.
5. J. VON NEUMANN, *Mathematische Grundlagen der Quantenmechanik*, Berlin, Julius Springer, 1932 (English translation, Princeton, Princeton University Press, 1955), Chapter VI. HEISENBERG (*Daedalus*, 87, 1958, p. 100), puts it even more concisely and picturesquely: 'The conception of objective reality . . . has thus evaporated . . . into the transparent clarity of a mathematics that represents no longer the behaviour of elementary particles but rather our knowledge of this behaviour.'
6. It is interesting from a psychological-epistemological point of view that, although consciousness is the only phenomenon for which we have direct evidence, many people deny its reality. The question: 'If all that

exists are some complicated chemical processes in your brain, why do you care what those processes are?' is countered with evasion. One is led to believe that, as explained in the text, the word 'reality' does not have the same meaning for all of us.

7. For this concept, see, e.g. E. WIGNER, *Gruppentheorie*, etc., Braunschweig, Friedr. Vieweg, 1931 (English translation, New York, Academic Press, Inc., 1959), Chapter XII.
8. VON NEUMANN, *op. cit.*, Chapter V
9. The only paper that is available on this subject seems to be 'The General and Logical Theory of Automata' in *The Hixon Symposium* (edited by L. A. Jeffress), New York, John Wiley and Sons, 1951, p.1. However, C. E. Shannon's discussion of von Neumann's work (*Bull. Amer. Math. Soc.*, 64, 1958, p. 123, draws attention to further unpublished papers. In connection with the reliability problem, to be discussed below, see 'Probabilistic Logics and the Synthesis of Reliable Organisms from Unreliable Components,' in *Automata Studies* (edited by C. E. Shannon and J. McCarthy), Princeton, Princeton University Press, 1956, p. 43.
10. Elsasser, op. cit., p. 129.
11. It must suffice to mention a few of the pertinent papers here and the knowledge of most of these I owe to Dr. H. Jehle of George Washington University: F. H. C. CRICK and J. D. WATSON, *Nature*, *171*, 1953, p. 737; *Proc. Roy. Soc. A 223*, 1954, p. 80; G. GAMOW, *Biol. Medd. Danske Vid. Selskab*, 22, 1954, No. 2; 22, 1955, No. 8; F. H. C. CRICK, J. S. GRIFFITH and L. E. ORGEL, *Proc. Nat. Acad. Sc.*, 43, 1957, p. 416; M. DELBRÜCK, S. W. GOLOMB, L. R. WELCH, *Biol. Medd. Danske Vid. Selskab*, 23, 1958, No. 9. See also H. J. MULLER, *Proc. Roy. Soc. B*. 134, 1947, p. 1.
12. P. MORRISON, *Amer. J. Physics*, 26, 1958, p. 358.

BIBLIOGRAPHY

Scientific Papers by Michael Polanyi

1910
1. M. Polanyi, 'Chemistry of the Hydrocephalic Liquid.' *Magyar ord. Archiv.*, N.F. *11*, p. 116 (1910).

1911
2. M. Polanyi, 'Investigation of the Physical and Chemical Changes of the Blood Serum During Starvation.' *Biochem. Z.*, *34*, p. 192 (1911).
3. M. Polanyi, 'Contribution to the Chemistry of the Hydrocephalic Liquid.' *Biochem. Z.*, *34*, p. 205 (1911).

1913
4. M. Polanyi and J. Baron, 'On the Application of the Second Law of Thermodynamics to Processes in the Animal Organism.' *Biochem. Z.*, *53*, p. 1 (1913).
5. M. Polanyi, 'A New Thermodynamic Consequence of the Quantum Hypothesis.' *Verh. deut, phys. Ges.*, *15*, p. 156 (1913).
6. M. Polanyi, 'New Thermodynamic Consequences of the Quantum Hypothesis.' *Z. phys. Chem.*, *83*, p. 339 (1913).

1914
7. M. Polanyi, 'Adsorption and Capillarity from the Standpoint of the Second Law of Thermodynamics.' *Z. phys. Chem.*, *88*, p. 622 (1914).
8. M. Polanyi 'Adsorption, Swelling and Osmotic Pressure of Colloids.' *Biochem. Z.*, *66*, p. 258 (1914).
9. M. Polanyi, 'On the derivation of Nernst's Theorem.' *Verh. deut. phys. Ges.*, *16*, p. 333 (1914).
10. M. Polanyi, 'On Adsorption from the Standpoint of the Third Law of Thermodynamics.' *Verh. deut, phys. Ges.*, *16*, p. 1012 (1914).

1915
11. M. Polanyi, 'On the Derivation of Nernst's Theorem.' *Verh. deut. phys. Ges. 17*, p. 350 (1915).

1916
12. M. Polanyi, 'Adsorption of Gases by a Solid Non-Volatile Adsorbent.' *Verh. deut. phys. Ges.*, *18*, p. 55 (1916).
13. M. Polanyi, 'New Procedure to Save Washing Materials.' Vegyeszeti Lapok 12 (1916).

1917
14. M. Polanyi, 'Adsorption of Gases by a Solid Non-Volatile Adsorbent.' *Ph. D. Thesis, Budapest* (1917).
15. M. Polanyi, 'On the Theory of Adsorption.' *Magyar Chem. Folyoirat*, *23*, p. 3 (1917).

1919
16. M. Polanyi and L. Mandoki, 'On the Causes of the Conductivity of Casein Solutions.' *Magyar Chem. Folyoirat*, *25*, (1919).

Bibliography

17. M. Polanyi, 'Conductivity-Lowering and Adsorption in Lyophilic Colloids.' *Magyar Chem. Folyoirat*, 25 (1919).

1920

18. M. Polanyi, 'Reaction Isochore and Reaction Velocity from the Standpoint of Statistics.' *Z. Elektrochem*, 26, p. 49 (1920).
19. M. Polanyi, 'On the Absolute Saturation of Attractive Forces Acting between Atoms and Molecules.' *Z. Elektrochem.*, 26, p. 261 (1920).
20. M. Polanyi, 'On the Problem of Reaction Velocity.' *Z. Elektrochem.*, 26, p. 228 (1920).
21. M. Polanyi, 'Correction to the Paper "Reaction Isochore and Reaction Velocity from the Standpoint of Statistics".' *Z. Elektrochem.*, 26, p. 231 (1920).
22. M. Polanyi, 'On Adsorption and the Origin of Adsorption Forces.' *Z. Elektrochem.*, 26, p. 370 (1920).
23. M. Polanyi, 'On the Nonmechanical Nature of Chemical Processes.' *Z. Physik*, 1, p. 337 (1920).
24. M. Polanyi, 'On the Theory of Reaction Velocity. *Z. Physik*, 2, p. 90 (1920).
25. M. Polanyi, 'Adsorption from Solutions of Substances of Limited Solubility.' *Z. Physik*, 2, p. 111 (1920).
26. M. Polanyi, 'On the Origin of Chemical Energy.' *Z. Physik*, 3, p. 31 (1920).
27. M. Polanyi, R. O. Herzog, and W. Jancke, 'X-Ray Spectroscopic Investigations on Cellulose', II. *Z. Physik*, 3, p. 343 (1920).
28. M. Polanyi, 'Studies on Conductivity-Lowering and Adsorption in Lyophilic Colloids'. *Biochem. Z.*, 104, p. 237 (1920).
29. M. Polanyi and L. Mandoki, 'The Origins of Conductivity in Casein Solutions.' *Biochem. Z.*, 104, p. 257 (1920).
30. M. Polanyi, 'Advances in the Theoretical Explanation of Adsorption.' *Chem. Ztg.*, 44, p. 340 (1920).

1921

31. M. Polanyi, 'On the Adsorption of Gases on Solid Substances.' *Festschr. Kaiser Wilhelm Ges. Zehnjahr. Jub.*, p. 171 (1921).
32. M. Polanyi, 'Fibrous Structure by X-Ray Diffraction.' *Naturwiss.*, 9, p. 337 (1921).
33. M. Polanyi, 'On the Current Resulting from the Compression of a Soldered Joint.' *Z. phys. Chem.*, 97, p. 459 (1921).
34. M. Polanyi, 'On Adsorption Catalysis.' *Z. Elektrochem.*, 27, p. 142 (1921).
35. M. Polanyi, E. Ettisch, and K. Weissenberg, 'Fibrous Structure of Hard-Drawn Metal Wires.' *Z. phys. Chem.*, 99, p. 332 (1921).
36. M. Polanyi, K. Becker, R. O. Herzog, and W. Jancke, 'On Methods for the Arrangement of Crystal Elements.' *Z. Physik*, 5, p. 61 (1921).
37. M. Polanyi, 'The X-Ray Fibre Diagram.' *Z. Physik*, 7, p. 149 (1921).
38. M. Polanyi, M. Ettisch, and K. Weissenberg, 'On Fibrous Structure in Metals.' *Z. Physik*, 7, p. 181 (1921).
39. M. Polanyi, 'On the Nature of the Tearing Process.' *Z. Physik*, 7, p. 323 (1921).
40. M. Polanyi, E. Ettisch, and K. Weissenberg, 'X-Ray Investigation of Metals.' *Physik. Z.*, 22, p. 646 (1921).

1922

41. M. Polanyi, 'The Reinforcement of Monocrystals by Mechanical Treatment.' *Z. Elektrochem.*, 28, p. 16 (1922).
42. M. Polanyi, 'Reflection on Mr. A. Eucken's Work: *On the Theory of Adsorption Processes*.' *Z.Elektrochem.*, 28, p. 110 (1922).
43. M. Polanyi, 'Determination of Crystal Arrangement by X-Ray Diffraction.' *Naturwiss.*, 10, p. 411 (1922).
44. M. Polanyi and K. Weissenberg, 'The X-Ray Fibre Diagram.' *Z. Physik*, 9, p. 123 (1922).
45. M. Polanyi and K. Weissenberg, 'The X-Ray Fibre Diagram.' *Z. Physik*, 10, p. 44 (1922).
46. M. Polanyi, H. Mark, and E. Schmid, 'Processes in the Stretching of Zinc

Bibliography

Crystals. I. General Description of the Phenomena and Research Methods.' *Z. Physik*, 12, p. 58 (1922).
47. M. Polanyi, H. Mark, and E. Schmid, 'Processes in the Stretching of Zinc Crystals. II. Quantitative Consideration of the Stretching Mechanism.' *Z. Physik*, 12, p. 78 (1922).
48. M. Polanyi, H. Mark, and E. Schmid, 'Processes in the Stretching of Zinc Crystals. III. Relationship between the Fibre Structure and Reinforcement.' *Z. Physik*, 12, p. 111 (1922).

1923

49. M. Polanyi and K. Weissenberg, 'Röntgenographic Investigations on Worked Metals.' *Z. Tech. Physik*, 4, p. 199 (1923).
50. M. Polanyi and E. Schmid, 'Discussion of the Sliding Friction Dependence on Pressure Normal to the Sliding Plane.' *Z. Physik*, 16, p. 336 (1923).
51. M. Polanyi, 'On Structural Changes in Metals through Cold Working.' *Z. Physik*, 17, p. 42 (1923).
52. M. Polanyi and H. Mark, 'Lattice Structure, Sliding Directions and Sliding Planes of White Tin.' *Z. Physik*, 18, p. 75 (1923).
53. M. Polanyi, R. O. Herzog, and W. Jancke, 'On the Structure of the Cellulose and Silk Fibre.' *Z. Physik*, 20, p. 413 (1923).
54. M. Polanyi and G. Masing, 'Cold Working and Reinforcement.' *Erg. exakt. Naturw.*, 2, p. 177 (1923).
55. M. Polanyi, 'Structural Analysis by Means of X-Rays.' *Physik. Z.*, 24, p. 407 (1923).
56. M. Polanyi, H. Mark, and E. Schmid, 'Investigations of Monocrystalline Wires of Tin.' *Naturwiss.*, 11, p. 256 (1923).

1924

57. M. Polanyi and H. Mark, 'Correction to the Paper *Lattice Structure, Sliding Directions and Sliding Planes of White Tin.*' *Z. Physik*, 22, p. 200 (1924).
58. M. Polanyi, E. Schiebold, and K. Weissenberg, 'On the Development of the Rotating Crystal Method.' *Z. Physik*, 23, p. 337 (1924).
59. M. Polanyi and E. Ewald, 'Plasticity and Strength of Rock Salt under Water.' *Z. Physik*, 28, p. 29 (1924).
60. M. Polanyi and G. Masing, 'On the Increase of Tensile Strength of Zinc by Cold-Working.' *Z. Physik*, 28, p. 169 (1924).
61. M. Polanyi, 'Osmotic Pressure, Pressure of Swelling, and Adsorption.' *Z. phys. Chem.*, 114, p. 387 (1924).
62. M. Polanyi and E. Schmid, 'On the Structure of Worked Metals.' *Z. tech. Physik*, 5, p. 580 (1924).
63. M. Polanyi and A. Schob, 'Stretching Experiments with Soft Vulcanized Rubber at the Temperature of Liquid Air.' *Mitt. Materialprüfungsamt*, 42, p. 22 (1924).

1925

64. M. Polanyi, 'Deformation of Monocrystals.' *Z. Krist.*, 61, p. 49 (1925).
65. M. Polanyi, 'Moulding of Solid Bodies from the Standpoint of Crystal Structure.' *Vortr. Dresden. Tag. Ges. angew. Math. Mech.*, 5, p. 125 (1925).
66. M. Polanyi, 'An Elongating Apparatus for Threads and Wires.' *Z. tech. Physik*, 6, p. 121 (1925).
67. M. Polanyi and E. Ewald, 'On the Form Strengthening of Rock Salt in Bending Experiments.' *Z. Physik*, 31, p. 139 (1925).
68. M. Polanyi and E. Ewald, 'Remarks on the Work of A. Joffe and M. Levitzky, On the Limits of Strength and Elasticity of Natural Rock Salt.' *Z. Physik*, 31, p. 746 (1925).
69. M. Polanyi and E. Schmid, 'Strengthening and Weakening of Sn Crystals.' *Z. Physik*, 32, p. 684 (1925).
70. M. Polanyi and E. Wigner, 'Formation and Decomposition of Molecules.' *Z. Physik*, 33, p. 429 (1925).

Bibliography

71. M. Polanyi and G. Sachs, 'On Elastic Hysteresis and Internal Strains in Bent Rock-Salt Crystals'. *Z. Physik*, *33*, p. 692 (1925).
72. M. Polanyi and M. Fischenich, 'The Origins of Conductivity in Casein Solutions.' *Kolloid-Z.*, *36*, p. 275 (1925).
73. M. Polanyi and H. Beutler, 'Chemiluminescence and Reaction Velocity.' *Naturwiss.*, *13*, p. 711 (1925).
74. M. Polanyi, 'Crystal Deformation and Strengthening.' *Z. Metallkunde*, *17*, p. 94 (1925).
75. M. Polanyi and G. Sachs, 'On the Release of Internal Strains by Annealing.' *Z. Metallkunde*, *17*, p. 227 (1925).

1926

76. M. Polanyi, H. Beutler, and S. v. Bogdandy, 'On Luminescence of Highly Dilute Flames.' *Naturwiss.*, *14*, p. 164 (1926).
77. M. Polanyi and S. v. Bogdandy, 'Ejection of Atoms from Solids by Chemical Attack on the Surface.' *Naturwiss.*, *14*, p. 1205 (1926).
78. M. Polanyi, 'Moulding of Metal Crystals, and the Moulded State.' *Werkstoff ausschuss Bericht No. 85*, p. 1 (1926).
79. M. Polanyi, S. v. Bogdandy and J. Boehm, 'On a Method of Producing *Molecular Mixtures*.' *Z. Physik*, *40*, p. 211 (1926).
80. M. Polanyi, 'Behaviour of Neutral Sodium Caseinogate in Membrane Hydrolysis.' *Biochem. Z.*, *171*, p. 473 (1926).
81. M. Polanyi and G. Sachs, 'Elastic Hysteresis in Rock Salt.' *Nature*, *116*, p. 692 (1926).

1927

82. M. Polanyi, R. L. Hasche, and E. Vogt, 'Spectral Intensity Distribution in the D-line of the Chemiluminescence of Sodium Vapour.' *Z. Physik*, *41*, p. 583 (1927).
83. M. Polanyi, 'The Structure of Matter and X-Ray Diffraction.' *Z. Ver. deut. Ing.*, *71*, p. 565 (1927).
84. M. Polanyi and S. v. Bogdandy, 'Rapid Analysis of Brass.' *Z. Metallkunde*, *19*, p. 164 (1927).
85. M. Polanyi and S. v. Bogdandy, 'Chemically-Induced Chain Reaction in Detonating Gas.' *Naturwiss.*, *15*, p. 410 (1927).
86. M. Polanyi, 'Theory of Wall Reactions.' *Chem. Rund. Mitteleuropa Balkan*, *4*, p. 160 (1927).
87. M. Polanyi and S. v. Bogdandy, 'Rapid Brass Analysis.' *Metal Ind.* (London), *30*, p. 195 (1927).
88. M. Polanyi and S. v. Bogdandy, 'Chemically Induced Chain Reactions in Mixtures of Halogens, Hydrogen and Methane.' *Z. Elektrochem.*, *33*, p. 554 (1927).

1928

89. M. Polanyi, 'Reply to the Letter of O. L. Sponster, Erroneous Determination of the Cellulose Space Lattice.' *Naturwiss.*, *16*, p. 263 (1928).
90. M. Polanyi, 'Deformation, Rupture and Hardening of Crystals.' *Naturwiss.*, *16*, p. 285 (1928).
91. M. Polanyi, 'Theoretical and Experimental Strength.' *Naturwiss.*, *16*, p. 1043 (1928).
92. M. Polanyi and F. Goldmann, 'Adsorption of Vapours on Carbon and the Thermal Dilation of the Interface.' *Z. phys. Chem.*, *132*, p. 321 (1928).
93. M. Polanyi and K. Welke, 'Adsorption, Heat of Adsorption and Character of Attachment between Small Amounts of Sulphur Dioxide and Carbon.' *Z. phys. Chem.*, *132*, p. 371 (1928).
94. M. Polanyi and W. Heyne, 'Adsorption from Solutions.' *Z. phys. Chem.*, *132*, p. 384 (1928).
95. M. Polanyi and L. Frommer, 'On Heterogeneous Elementary Reactions. I. Action of Chlorine on Copper.' *Z. phys. Chem.*, *137*, p. 201 (1928).
96. M. Polanyi, 'Application of Langmuir's Theory to the Adsorption of Gases on Charcoal.' *Z. phys. Chem. A 138*, p. 459 (1928).

Bibliography

97. M. Polanyi and E. Wigner, 'On the Interference of Characteristic Vibrations as the Cause of Energy Fluctuations and Chemical Changes.' *Z. phys. Chem. A 139*, p. 439 (1928).
98. M. Polanyi and H. Beutler, 'On Highly Dilute Flames. I.' *Z. phys. Chem. B 1*, p. 3 (1928).
99. M. Polanyi and S. v. Bogdandy, 'On Highly Dilute Flames. II. Nozzle Flames. Increase of Light Emission With Increasing Partial Pressure of Sodium Vapour.' *Z. phys. Chem. B 1*, p. 21 (1928).
100. M. Polanyi and G. Schay, 'On Highly Dilute Flames. III. Sodium-Chlorine Flame. Evidence for and Analysis of the Reaction and Luminescence Mechanism. Both Reaction Types. Survey of the Whole Work.' *Z. phys. Chem.*, *B 1*, p. 30 (1928).
101. M. Polanyi and G. Schay, 'Correction to the Work. On Highly Dilute Flames. III.' *Z. phys. Chem.*, *B 1*, p. 384 (1928).
102. M. Polanyi, 'On the Simplest Chemical Reactions.' *Réunion Intern. Chim., Phys.*, p. 198 (1928).
103. M. Polanyi and H. Beutler, 'On Highly Dilute Flames, I.' *Z. Physik*, 47, p. 379 (1928).
104. M. Polanyi and G. Schay, 'Chemiluminescence between Alkali Metal Vapours and Tin Halides.' *Z. Physik*, 47, p. 814 (1928).
105. M. Polanyi, 'Deformation, Rupture and Hardening of Crystals.' *Trans. Faraday Soc.*, 24, p. 72 (1928).
106. M. Polanyi, 'The Inhibition of Chain Reactions by Bromine.' *Trans. Faraday Soc.*, 24, p. 606 (1928).

1929

107. M. Polanyi, 'Principles of the Potential Theory of Adsorption.' *Z. Elektrochem.*, 35, p. 431 (1929).
108. M. Polanyi, 'Consideration of Activation Processes at Surfaces.' *Z. Elektrochem.*, 35, p. 561 (1929).
109. M. Polanyi and E. Schmid, 'Problems of Plasticity. Deformation at Low Temperatures.' *Naturwiss.*, 17, p. 301 (1929).

1930

110. M. Polanyi, 'On the Nature of the Solid State.' *Metallwirt.*, 9, p. 553 (1930).
111. M. Polanyi and L. Frommer, 'On Gas Phase Luminescence in a Heterogeneous Reaction.' *Z. phys. Chem.*, *B 6*, p. 371 (1930).
112. M. Polanyi and H. v. Hartel, 'On Atomic Reactions Possessing Inertia.' *Z. phys. Chem.*, *B 11*, p. 97 (1930).
113. M. Polanyi, W. Meissner, and E. Schmid, 'Measurements with the Aid of Liquid Helium. XII. Plasticity of Metal Crystals at Low Temperatures.' *Z. Physik*, 66, p. 477 (1930).
114. M. Polanyi and H. Eyring, 'On the Calculation of the Energy of Activation.' *Naturwiss.*, 18, p. 914 (1930).
115. M. Polanyi and F. London, 'The Theoretical Interpretation of Adsorption Forces.' *Naturwiss.*, 18, p. 1099 (1930).
116. M. Polanyi and E. Schmid, 'Problems of Plasticity. Deformation at Low Temperatures.' *Mitt. deut. Materialprüfungs Anst. Sonderheft*, 10, p. 101 (1930).
117. M. Polanyi, 'The Nature of the Solid State.' *Umschau*, 34, p. 1001 (1930). *Mitt. deut. Materialprüfungs Anst. Sonderheft*, 13, p. 113 (1930).

1931

118. M. Polanyi and H. Eyring, 'On Simple Gas Reactions.' *Z. phys. Chem. B 12*, p. 279 (1931).
119. M. Polanyi and E. Cremer, 'Estimation of Molecular Lattice Dimensions from Resonance Forces.' *Z. phys. Chem.*, *B 14*, p. 435 (1931).
120. M. Polanyi and E. Cremer, 'Decrease of Fundamental Frequency as the First Stage of Chemical Reaction.' *Z. phys. Chem. Bodenstein Festband*, p. 720 (1931).

Bibliography

121. M. Polanyi and P. Beck, 'Recovery of Recrystallising Ability by Reformation.' *Z. Elektrochem.*, *37*, p. 521 (1931).
122. M. Polanyi and P. Beck, 'Recovery of Recrystallising Power by Reformation.' *Naturwiss.*, *19*, p. 505 (1931).
123. M. Polanyi, 'Atomic Reactions.' *Z. angew. Chem.*, *44*, p. 597 (1931).

1932

124. M. Polanyi and H. Ekstein, 'Note on the Mechanism of the Reaction $H_2 + I_2 \rightarrow 2\ HI$ and of Similar Reactions at Surfaces.' *Z. phys. Chem.*, *B 15*, p. 334 (1932).
125. M. Polanyi, E. Horn, and H. Sattler, 'On Highly Dilute Flames of Sodium Vapour with Cadmium Halides and Zinc Chloride'. *Z. phys. Chem.*, *B 17*, p. 220 (1932).
126. M. Polanyi, H. v. Hartel and N. Meer, 'Investigation of the Reaction Velocity between Sodium Vapour and Alkyl Chlorides.' *Z. phys. Chem.*, *B 19*, p. 139 (1932).
127. M. Polanyi and N. Meer, 'Comparison of the Reactions of Sodium Vapour with Other Organic Processes.' *Z. phys. Chem.*, *B 19*, p. 164 (1932).
128. M. Polanyi and E. Cremer, 'Test of the "Tunnel" Theory of Heterogeneous Catalysis; the Hydrogenation of Styrene.' *Z. phys. Chem.*, *B 19*, p. 443 (1932).
129. M. Polanyi, 'Developments in the Theory of Chemical Reactions.' *Naturwiss.*, *20*, p. 289 (1932).
130. M. Polanyi and D. W. G. Style, 'On an Active Product of the Reaction between Sodium Vapour and Alkyl Halides.' *Naturwiss.*, *20*, p. 401 (1932).
131. M. Polanyi, *Atomic Reactions*. William & Norgate Ltd., London (1932).
132. M. Polanyi, 'Theories of the Adsorption of Gases. A General Survey and Some General Remarks.' *Trans. Faraday Soc.*, *28*, p. 316 (1932).
133. M. Polanyi, 'The Theory of Chemical Reactions.' *Uspekhi Khim.*, *1*, p. 345 (1932).

1933

134. M. Polanyi, S. v. Bogdandy and G. Veszi, 'On a Method for the Preparation of Colloids and for Hydrogenation with Atomic Hydrogen.' *Angew. Chem.*, *46*, p. 15 (1933). *Chem. Fabrik*, *6*, p. 1 (1933).
135. M. Polanyi and E. S. Gilfillan, 'Micropycnometre for the Determination of Displacements of Isotopic Ratio in Water.' *Z. phys. Chem.*, *A 166*, p. 254 (1933).
136. M. Polanyi, E. Bergmann, and A. Szabo, 'The Mechanism of Simple Substitution Reactions and the Walden Inversion.' *Z. phys. Chem.*, *B 20*, p. 161 (1933).
137. M. Polanyi and J. Curry, 'On the Reaction between Sodium Vapour and Cyanogen Halides.' *Z. phys. Chem.*, *B 20*, p. 276 (1933).
138. M. Polanyi and E. Cremer, 'The Conversion of o- into p-Hydrogen in the Solid State.' *Z. phys. Chem.*, *B 21*, p. 459 (1933).
139. M. Polanyi, E. Horn, and D. W. G. Style, 'On the Isolation of Free Methyl and Ethyl by the Reaction between Sodium Vapour and Methyl and Ethyl Bromides.' *Z. phys. Chem.*, *B 23*, p. 291 (1933).
140. M. Polanyi, E. Cremer, and J. Curry, 'On a Method for the Determination of the Velocity of Gaseous Reactions of Atomic Hydrogen.' *Z. phys. Chem.*, *B 23*, p. 445 (1933).
141. M. Polanyi, 'A Note on the Electrolytic Separation of Heavy Hydrogen by the Method of G. N. Lewis.' *Naturwiss.*, *21*, p. 316 (1933).
142. M. Polanyi and E. Bergmann, 'Autoracemization, and Velocity of Electrolytic Dissociation.' *Naturwiss.*, *21*, p. 378 (1933).
143. M. Polanyi, 'Adsorption and Capillary Condensation.' *Phys. Z. Sowjetunion*, *4*, p. 144 (1933).
144. M. Polanyi, 'A Method for the Measurement of Gaseous Reactions.' *Nature* **132**, p. 747 (1933).

Bibliography

145. M. Polanyi and J. Horiuti, 'A Catalysed Reaction of Hydrogen with Water.' *Nature*, *132*, p. 819 (1933).
146. M. Polanyi and J. Horiuti, 'Catalyzed Reaction of Hydrogen with Water, and the Nature of Over-voltage.' *Nature*, *132*, p. 931 (1933).
147. M. Polanyi, 'Atomic Reactions.' *Uspekhi Khim.*, *2*, p. 412 (1933).

1934

148. M. Polanyi and E. Horn, 'On the Isolation of Free Phenyl Radicals by the Reaction of Sodium Vapour with Bromobenzene.' *Z. phys. Chem.*, B 25, p. 151 (1934).
149. M. Polanyi, E. Horn, and D. W. G. Style, 'The Isolation of Free Methyl and Ethyl by the Reaction between Sodium Vapour and Methyl and Ethyl Bromides.' *Trans. Faraday Soc.*, *30*, p. 189 (1934).
150. M. Polanyi and A. L. Szabo, 'On the Mechanism of Hydrolysis. The Alkaline Saponification of Amyl Acetate.' *Trans. Faraday Soc.*, *30*, p. 508 (1934).
151. M. Polanyi and L. Frommer, 'A New Method for Measuring the Rate of High Velocity Gas Reactions.' *Trans. Faraday Soc.*, *30*, p. 519 (1934).
152. M. Polanyi, J. Horiuti, and G. Ogden, 'Catalytic Replacement of Haplogen by Diplogen in Benzene.' *Trans. Faraday Soc.*, *30*, p. 663 (1934).
153. M. Polanyi and J. Horiuti, 'Exchange Reaction of Hydrogen on Metal Catalysts.' *Trans. Faraday Soc.*, *30*, p. 1164 (1934).
154. M. Polanyi and R. A. Ogg, Jr., 'The Mechanism of Ionogenic Reactions.' *Mem. Proc. Manchester Lit. Phil. Soc.*, *78*, p. 41 (1934).
155. M. Polanyi and J. Horiuti, 'On the Mechanism of Ionisation of Hydrogen at a Platinum Electrode.' *Mem. Proc. Manchester Lit. Phil.Soc.*, *78*, p. 47 (1934).
156. M. Polanyi, 'On a Form of Lattice Distortion that May Render a Crystal Plastic.' *Z. Physik*, *89*, p. 660 (1934).
157. M. Polanyi, 'Reaction Rates of the Hydrogen Isotopes.' *Nature*, *133*, p. 26 (1934).
158. M. Polanyi and J. Horiuti, 'Catalytic Hydrogen Replacement, and the Nature of Over-voltage.' *Nature*, *133*, p. 142 (1934).
159. M. Polanyi, B. Cavanagh, and J. Horiuti, 'Enzyme Catalysis of the Ionisation of Hydrogen.' *Nature*, *133*, p. 797 (1934).
160. M. Polanyi and J. Horiuti, 'Catalytic Interchange of Hydrogen between Water and Ethylene and between Water and Benzene.' *Nature*, *134*, p. 377 (1934).
161. M. Polanyi and J. Horiuti, 'Direct Introduction of Deuterium into Benzene.' *Nature*, *134*, p. 847 (1934).
162. M. Polanyi, R. A. Ogg, Jr., and L. Werner, 'Optical Inversion by Negative Substitution.' *Chem. and Ind.*, *53*, p. 614 (1934).
163. M. Polanyi, 'Discussion on Heavy Hydrogen.' *Proc. Roy. Soc. (London)*, A 144, p. 14 (1934).
164. M. Polanyi, 'Discussion on Energy Distribution in Molecules.' *Proc. Roy. Soc. (London)*, A 146, p. 253 (1934).
165. M. Polanyi and W. Heller, 'Quantitative Studies of Atomic Reactions.' *Compt. rend.*, *199*, p. 1118 (1934).
166. M. Polanyi, 'Discussion of Methods of Measuring and Factors Determining the Speed of Chemical Reactions.' *Proc. Roy. Soc. (London)*, B 116, p. 202 (1934).

1935

167. M. Polanyi and R. A. Ogg, Jr., 'Substitution of Free Atoms and Walden Inversion. The Decomposition and Racemisation of Optically Active sec-Butyl Iodide in the Gaseous State.' *Trans. Faraday Soc.*, *31*, p. 482 (1935).
168. M. Polanyi and R. A. Ogg, Jr., 'Mechanism of Ionic Reactions.' *Trans. Faraday Soc.*, *31*, p. 604 (1935).
169. M. Polanyi and M. G. Evans, 'Some Applications of the Transition State Method to the Calculation of Reaction Velocities, Especially in Solution.' *Trans. Faraday Soc.*, *31*, p. 875 (1935).
170. M. Polanyi and R. A. Ogg, Jr., 'Diabatic Reactions and Primary Chemiluminescence.' *Trans. Faraday Soc.*, *31*, p. 1375 (1935).

Bibliography

171. M. Polanyi, 'Heavy Water in Chemistry.' *Nature,* 135, p. 19 (1935).
172. M. Polanyi, J. Kenner, and P. Szego, 'Aluminium Chloride as a Catalyst of Hydrogen Interchange.' *Nature,* 135, p. 267 (1935).
173. M. Polanyi, G. H. Bottomley and B. Cavanagh, 'Enzyme Catalysis of the Exchange of Deuterium with Water.' *Nature,* 136, p. 103 (1935).
174. M. Polanyi, 'Adsorption and Catalysis.' *J. Soc. Chem. Ind.,* 54, p. 123 (1935).
175. M. Polanyi, 'Heavy Water.' *J.Soc. Dyers Colourists,* 51, p. 90 (1935).
176. M. Polanyi and J. Horiuti, 'Principles of a Theory of Proton Transfer.' *Acta Physicochim. U.S.S.R.,* 2, p. 505 (1935).

1936

177. M. Polanyi and W. Heller, 'Reactions between Sodium Vapour and Volatile Polyhalides, Velocities and Luminescence.' *Trans. Faraday Soc.,* 32, p. 663 (1936).
178. M. Polanyi, E. Bergmann, and A. L. Szabo, 'Substitution and Inversion of Configuration.' *Trans. Faraday Soc.,* 32, p. 843 (1936).
179. M. Polanyi and M. G. Evans, 'Further Considerations on the Thermodynamics of Chemical Equilibria and Reaction Rates.' *Trans. Faraday Soc.* 32, p. 1333 (1936).
180. M. Polanyi and D. D. Eley, 'Catalytic Interchange of Hydrogen with Water and Alcohol.' *Trans. Faraday Soc.,* 32, p. 1388 (1936).
181. M. Polanyi, M. Calvin, and E. G. Cockbain, 'Activation of Hydrogen by Phthalocyanine and Copper Phthalocyanine.' I. *Trans. Faraday Soc.,* 32, p. 1436 (1936).
182. M. Polanyi, M. Calvin, and D. D. Eley, 'Activation of Hydrogen by Phthalocyanine and Copper Phthalocyanine. II.' *Trans. Faraday Soc.,* 32, p. 1443 (1936).
183. M. Polanyi and M. G. Evans, 'Equilibrium Constants and Velocity Constants.' *Nature,* 157, p. 530 (1936).
184. M. Polanyi and C. Horrex, 'Atomic Interchange between Water and Saturated Hydrocarbons.' *Mem. Proc. Manchester Lit. Phil. Soc.,* 80, p. 33 (1936).

1937

185. M. Polanyi and M. G. Evans, 'On the Introduction of Thermodynamical Variables into Reaction Kinetics.' *Trans. Faraday Soc.,* 33, p. 448 (1937).
186. M. Polanyi, 'The Transition State in Chemical Reactions.' *J. Chem. Soc.,* p. 629 (1937).
187. M. Polanyi, 'The Transition State in Chemical Kinetics.' *Nature,* 139, p. 575 (1937).
188. M. Polanyi, 'Catalytic Activation of Hydrogen.' *Sci. J. Roy. Coll. Sci.,* 7, p. 21 (1937).
189. M. Polanyi, 'Colours as Catalysts.' *J. Oil Col. Chem. Assoc.,* Buxton Conf. No. 3 (1937).

1938

190. M. Polanyi and M. G. Evans, 'Inertia and Driving Force of Chemical Reactions.' *Trans. Faraday Soc.,* 34, p. 11 (1938).
191. M. Polanyi, 'On the Catalytic Properties of Phthalocyanine Crystals.' *Trans. Faraday Soc.,* 34, p. 1191 (1938).
192. M. Polanyi, 'The Deformation of Solids.' Report Reunion Int. Phys. Chim. Biol., (1938).
193. M. Polanyi, P. Debye, F. Simon, M. Wiersma, C. V. Raman, and B. van der Pol, *General Physics,* Hermann & Cie, Paris (1938).

1939

194. M. Polanyi and M. G. Evans, 'Notes on the Luminescence of Sodium Vapour in Highly Dilute Flames.' *Trans. Faraday Soc.,* 35, p. 178 (1939).
195. M. Polanyi, C. Horrex, and R. K. Greenhalgh, 'Catalytic Exchange of Hydrogen.' *Trans. Faraday Soc.,* 35, p. 511 (1939).
196. M. Polanyi and R. K. Greenhalgh, 'Hydrogenation and Atomic Exchange of Benzene.' *Trans. Faraday Soc.,* 35, p. 520 (1939).

Bibliography

1940

197. M. Polanyi and A. R. Bennett, 'Influence of Acidity on Catalytic Exchange of Hydrogen and Water.' *Trans. Faraday Soc.*, 36, p. 377 (1940).
198. M. Polanyi and E. T. Butler, 'Influence of Substitution on Organic Bond Strength.' *Nature*, 146, p. 129 (1940).
199. M. Polanyi and E. C. Baughan, 'Energy of Aliphatic Carbon Linking.' *Nature*, 146, p. 685 (1940).

1941

200. M. Polanyi, E. C. Baughan, and M. G. Evans, 'Covalency, Ionisation and Resonance in Carbon Bonds.' *Trans. Faraday Soc.*, 37, p. 377 (1941).
201. M. Polanyi and E. C. Baughan, 'Activation Energy of Ionic Substitution.' *Trans. Faraday Soc.*, 37, p. 648 (1941).
202. M. Polanyi and M. G. Evans, 'Effect of Negative Groups on Reactivity.' *Nature*, 148, p. 436 (1941).

1942

203. M. Polanyi and A. G. Evans, 'Calculation of Steric Hindrance.' *Nature*, 149, p. 608 (1942).

1943

204. M. Polanyi and E. T. Butler, 'Rates of Pyrolysis and Bond Energies of Substituted Organic Iodides, I.' *Trans. Faraday Soc.*, 39, p. 19 (1943).
205. M. Polanyi, 'Resonance and Chemical Reactivity.' *Nature*, 151, p. 96 (1943).
206. M. Polanyi and A. G. Evans, 'Steric Hindrance and Heats of Formation.' *Nature*, 152, p. 738 (1943).

1945

207. M. Polanyi, E. T. Butler, and E. Mandel, 'Rates of Pyrolysis and Bond Energies of Substituted Organic Iodides. II.' *Trans. Faraday Soc.*, 41, p. 298 (1945).

1946

208. M. Polanyi, A. G. Evans, D. Holden, P. H. Plesch, H. A. Skinner, and M. A. Weinberger, 'Friedel-Crafts Catalysts and Polymerization.' *Nature*, 157, p. 102 (1946).
209. M. Polanyi, 'Activation of Catalysts in Olefine Reactions.' *Nature*, 157, p. 520 (1946).
210. M. Polanyi, A. G. Evans, and G. W. Meadows, 'Friedel-Crafts Catalysts and Polymerization.' *Nature*, 158, p. 94 (1946).

1947

211. M. Polanyi and A. G. Evans, 'Polymerization of iso-Butene by Friedel-Crafts Catalysts.' *J. Chem. Soc.*, 1947, p. 252.
212. M. Polanyi, P. H. Plesch, and H. A. Skinner, 'The Low Temperature Polymerization of iso-Butene by Friedel-Crafts Catalysts.' *J. Chem. Soc.*, 1947, p. 257.
213. M. Polanyi, A. G. Evans, and M. G. Evans, 'Mechanism of Substitution at a Saturated Carbon Atom.' *J. Chem. Soc.*, 1947, p. 558.
214. M. Polanyi, A. G. Evans, and G. W. Meadows, 'Friedel-Crafts Catalysts and Polymerization.' *Rubber Chem. and Technol.*, 20, p. 375 (1947).
215. M. Polanyi, A. G. Evans, and G. W. Meadows, 'Polymerization of Olefines by Friedel-Crafts Catalysts.' *Nature*, 160, p. 869 (1947).

1948

216. M. Polanyi, 'Polymerization at Low Temperatures'. *Angew. Chem.*, 60 A, p. 76 (1948).

1949

217. M. Polanyi, 'Mechanism of Chemical Reactions.' *Endeavour*, 8, p. 3 (1949).
218. M. Polanyi, 'Experimental Proofs of Hyperconjugation.' *J. Chim. Phys.*, 46, p. 235 (1949).

Bibliography

Books by Michael Polanyi

Atomic Reactions, London, Williams and Norgate, 1933.
USSR Economics, Manchester, Manchester University Press, 1935.
The Contempt of Freedom, London, Watts, 1940.
Full Employment and Free Trade, Cambridge, Cambridge University Press, 1945.
Science, Faith and Society, London, Oxford University Press, 1946; and New York.
The Logic of Liberty, London, Routledge and Kegan Paul, 1951; Chicago, Chicago University Press.
Personal Knowledge, London, Routledge and Kegan Paul, 1958; Chicago, Chicago University Press.
The Study of Man, London, Routledge and Kegan Paul, 1959; Chicago, Chicago University Press.
Beyond Nihilism, London, Cambridge University Press, 1960.

Film by Michael Polanyi

Unemployment and Money—A diagrammatic film prepared with the assistance of Miss Mary Field, Mr. R. Jeffryes and Professor J. Jewkes. (A description and theoretical analysis of this film is to be found in the *Review of Economic Studies*, Vol. *8*, p. 1, 1940.)

A partial list of journals in which articles by M. Polanyi have appeared

Advancement of Science.
Archiv der Staatswissenschaften.
Atti del Congresso di Metodologia.
British Journal for the Philosophy of Science.
Bulletin of Atomic Scientists.
Cambridge Journal.
Dialectica.
Economica.
Encounter.
Humanitas.
The Lancet.
The Listener.
The Manchester Guardian.
The Manchester School.
The Manchester Statistical Society.
Measure.
Memoirs and Proceedings of the Manchester Literary and Philosophical Society.
Nineteenth Century (*Twentieth Century*).
Physical Science and Human Values (Princeton 1947).
Political Quarterly.
Review of Economic Studies.
Science.
Scientific Monthly.
Symposium on Observation and Interpretation (Bristol 1957).
Time and Tide.
Zeitschrift fur Physikalische Chemie.